Cavemen with Cell Phones

BY
SIMON GRAY

safehouse books 

a division of
shinyblack.inc
707 south 23rd avenue east
duluth, mn 55812

CAVEMEN WITH CELL PHONES

made on a mac.

first edition

ISBN 0-9744912-0-9

CONTENTS

"We are what we repeatedly do."

Aristotle

Cavemen With Cell Phones?

Isn't that exactly what we are? Roaming around the planet in our SUVs, hunting and gathering and fighting like it's still 50,000 B.C. Have we really changed all that much from our stoop-shouldered grunting ancestors? Or are we still doing the same things we've always done, except now we have newer tools?

Birds fly south for the winter. Fish swim upstream to spawn. These are instinctual behaviors. Some people will shop at bargain stores and buy 100 rolls of toilet paper and a gallon jug of ketchup. Is this instinctual behavior? Is it Gathering Instinct, motivated out of the fear of running out?

We are animals.

And, our behavior is more similar to the behavior of other animals than it is different.

We eat, sleep, drink, converse, have sex, entertain ourselves, raise our young, poop, and pee. Everything we create, invent, design, and build is done to enhance our experiences of eating, sleeping, drinking, conversing, having sex, entertaining ourselves, raising our young, pooping, and peeing. Our motivations haven't changed even though the way we do things changes. We don't live in caves anymore, but we still act like cavemen.

Cavemen

Anthropologists estimate that we human beings have existed on the earth for at least three and a half million years. For most of that time our average life expectancy was between 20 and 25 years. Up to that age, these Instincts seem to perform well for us. Beyond that age, however, they seem to cause problems. This could be due to the fact that we've never lived this long before, and there are many more of us trying to share the same space and resources. As Primitive people, we lived in small communities and spent most of our time taking care of our basic needs. We hunted and gathered our own food, built or sought out shelter, had children, raised and protected our young, and defended ourselves against other humans and animals that threatened our safety. Most of this was done in fields or in caves, in any kind of weather, and with whatever crude tools we could fashion.

WITH CELL PHONES

In the last ten thousand years or so, we human being have become "civilized" , and we have evolved into a different kind of creature. Now, many of us no longer depend directly on the earth for our needs. Rather, we depend on other people for our needs. Other people build our shelters, other people provide our food, and other people insure our safety. There are a lot more of us than there ever has been, and we live three to four times longer than we ever have.

We are simple creatures. We think we are complex, but that's because we don't understand why we behave the way we do. And, we don't understand why other people behave they way they do. We have invented several ways to try to explain behavior: We use psychology, biology, religion, and astrology.

But there may be another explanation: Instinct. We may not be too different from our cavemen ancestors after all. Our behaviors, even though they can take different forms, might still be motivated out of the same basic primitive Instincts that motivated our ancestors. **How** we do things changes. We continually improve technology. One human being anywhere in the world can talk to another human being anywhere in the world. But **why** we do things hasn't changed in almost four million years. We may now have cell phones, but we still act like cavemen.

CIVILIZATION

Civilization is the idea that we can organize eating, sleeping, having sex, raising our young, pooping, and peeing. But how do we organize this stuff? We need cooperation.

Civilized nations exist because the people who live in them agree that the nation exists. Nations are not based on certain pieces of land, as borders have been constantly redrawn for centuries. Old nations die and new ones are formed. Nations are not based on race. All the Chinese people in the world don't live in China. If you see a few Chinese people living in another country they may band together in a community. In some towns there are neighborhoods called "China towns" where Chinese people seem to congregate to live and work together. They do this because there is comfort, safety, communication and shared heritage that bonds them. But these things do not bond them with non-Chinese people in the same city.

Nations are nations by the agreement of their citizens. The way they agree is by following the rules and laws of a nation, and in return they enjoy the benefits of security, protection, dependable food and water supply, and shelter. What keeps people of different races and religions living and working together as a nation is money. As long as the currency is good, life goes on. People work and spend their paychecks, which supports business and trade, and so the whole nation stays together.

If we woke up tomorrow morning and the stock market had collapsed, and our currency had no value, our first thoughts would probably be for the basic needs of ourselves and our families: food, shelter, and safety. We would not be able to buy food, so we would have to find some other way to get it. And, other people might try to take away our food or shelter, and so we would need to protect ourselves against them with some form of weapon. We would immediately, and instinctively, start acting like our primitive ancestors.

Money is the glue of a civilization. If the money is no good, the next thing used to hold a nation together is usually force and weapons. But if you can't pay the soldiers, the whole thing goes down the toilet. In essence, we would become cave people again in the blink of an eye. And the survival Instincts we would need are already inside of us.

So if these Instincts are already there, then what effect are they having on our day-to-day lives?

CAVEMAN INSTINCTS

Like the other animals, we act on basic Instincts. These Instincts motivate us to do certain behaviors that have served to insure our survival. These Instincts have nothing to do with rational thought or common sense. They are knee-jerk reactions to common situations. They can be classified into eight specific types:

HUNTING
GATHERING
WARRIOR
WORKER
INVENTOR
And the Breeding Instincts
ATTRACTION
MATING
and **NURTURING**

The first five Instincts serve to keep us alive as a species. The three Breeding Instincts serve to keep us reproducing. There could still be other types not yet identified, but these eight seem to be the most obvious.

Some of our Instincts seem to be more active in some people and less active in others. Some of us seem to act primarily out of one Instinct. Some of us seem to be motivated by a combination such as Hunting and Gathering. Some of us seem to have an equal balance where no single Instinct is dominant. We may be motivated more out of one Instinct during certain times of the day, or times of the month (try sitting in an emergency room or a police station during a full moon), or times of the year, or periods in our lives.

These are the Instincts which have kept us alive as a species for almost 4 million years.

BIG PICTURE & SMALL PICTURE

There are major differences in the way our Instincts think and behave. The older we get, and the more technologically advanced we become, the more we realize that we are able to manipulate things towards satisfying our desires. Each one of our Instincts has its own agenda. It sees the world in a very specific way and has an established set of beliefs that it acts out of. These beliefs are not universal to all of our Instincts. And so, miscommunication can occur.

Our Inventor, Hunting, and Worker Instincts see the big picture of reality. Those of us with these Instincts see patterns and trends. This enables us to be effective in our roles in society. But, we often miss the small picture. As those of us who have Inventors for partners can tell you, they may have figured out a way to make the Internet child safe, but they have no idea what water temperature to wash their shirts in. Our Warrior, Gathering, and the Breeding Instincts tend to see a much narrower picture of reality. Those of us with these Instincts stay focused on the issues at hand. Likewise, we can often miss the big picture entirely.

And so we spend a great deal of time puzzled and critical of others because they don't see things the way we do. It is a question of focus: Warrior, Gathering, and the Breeding Instincts are focused on the here and now, on today, the task at hand, the immediate issue. Inventor, Hunter, and Worker Instincts are focused on the long term and what has been tried before. All of our Instincts are looking in different directions.

These eight Instincts also tend to see different degrees of the big picture and the small picture. Warrior Instinct sees the narrowest view of all. It sees black and white, right or wrong, and no in-between. Its view is rigid and forceful. The Breeding Instincts are a little less narrow than Warrior Instinct. For instance; Nurturing Instinct sees good and bad. And good and bad, unlike right and wrong, can change depending on new information. Gathering Instinct tends to see an even slightly wider view than Nurturing. This allows us to continually include new things into our gathering. On the other hand, our Worker Instinct sees a much wider view than our Gathering Instinct. We are the ones who take everything with a grain of salt, don't panic or obsess about anything, and sleep well at night. Our Hunting Instinct sees a slightly larger picture than that. It sees that we can modify and improve the way we live, and make it more efficient and effective. Our Inventor Instinct sees an even broader view than Hunting. We Inventors dream up new technologies to improve societies. Our Inventor Instinct sees the biggest picture of all.

Imagine 8 people: 3 blind and 5 deaf, all touching different parts of the same elephant, and each one of us thinking that we have the whole thing figured out, and trying to communicate with each other.

Our miscommunication occurs because we think other people **should** see things the way we do. We think they **should** think and act the way we think and act. Our Warrior Instinct sees people with Gathering Instinct as slow and stupid. Our Inventor Instinct sees Warrior Instinct as narrow-minded and stupid. Nurturing see Inventor as negative and stupid. Nurturing Instinct shoulds on Worker Instinct. Gathering Instinct shoulds on Warrior. Warrior shoulds on everyone. They don't speak our language and we don't speak theirs. Our own Instincts serve us so well, we believe they can serve others too, so we try to get them to think and act like we do. The only problem is: everyone is thinking something different! None of our Instincts speak the same language. It is like yelling at bees.

This is not a judgment. The big picture is not **better** that the small picture. The small picture is not **better** than the big picture. One Instinct is not better or worse than another. One is not good and the other bad. One is not right and the other wrong. A house is built of wood, bricks, glass, and metal. We don't shame wood for not being bricks. We don't use glass where we need to use metal. We need all of them to build a house. We apparently need all of these Instincts to keep us alive. All are needed, and all are just as valuable in their particular roles. And, since we seem to have more than one Instinct inside us, we can end up with a combination that allows us to see everything, and balance our behavior accordingly.

When reading about these Instincts it is very important to remember :

⏩ These are not judgments! The point is not to judge our Instincts as being bad or good, or wrong or right. The point is to try to identify the motivation behind our behaviors.

⏩ Instincts have nothing to do with rational thought or common sense. Our behaviors are knee-jerk reactions to situations and comments. They are not objective rational choices of action.

- Remember -

All of us have **all** of these Instincts inside us. We don't just have Worker, or Warrior Instinct, we have combinations of all eight Instincts. Some may be more obvious than others. Some may be more active than others. All people are not the same. All people with the same Instincts are not the same. Nurturing Instinct believes in absolutes: all, always, never, only, every. But even all people with Nurturing Instinct can't agree on what children should learn, and how and when.

At the end of each chapter there is a section on partners. As you will see, our Instincts can often choose our partners for us more often than our rational thinking. We can be attracted to other people for reasons which may not always be obvious to us.

THIS IS A THEORY

There is no proof that the ideas in this book are true. There is no documented evidence to support any of the claims made here. Likewise, there is no proof that certain behaviors are **not** motivated by primitive Instincts. All we can do is observe patterns of behavior and draw conclusions. That is the basis for modern Psychology, and that is the basis for this book.

Behaviors seem to be either learned or forced or instinctual. We can learn new ways to act in certain situations. We can force ourselves, or try to force others, to act in certain ways. But there are also certain behaviors that are identical among groups of people who do not share the same history, country, or generation. That is the focus of this book.

This is not a Personality Index like the astrological chart of the zodiac signs. It is not a psychological index like the Myers-Briggs or the Minnesota Multi-phasic Personality Inventory. It is a **Behavior Index** based on the observation of repeated similar behaviors in different people, and tracing the motivation for those behaviors.

Either the information in this book will ring true for you, or it won't. I'm no psychologist, but I am the world's foremost expert on me. Nobody knows more about me than I do, and I know I have **all** of these Instincts. How many do you identify with? Are you a caveman too?

caveman

instincts

Hunting Instinct

Hunting Instinct is concerned with details, research, efficiency, numbers, facts, and proof. Those of us with Hunting Instinct seek the best way, the most efficient way to get something done. At first glance, our Hunting and Warrior Instincts may seem similar. There are significant differences. Our Hunting Instinct sees the big picture, and our Warrior Instinct sees the small picture. Those of us with Hunting Instinct don't fight with other people; to us it is an inefficient means of conflict resolution. We are efficiency experts; we calculate and plan. Where our Warrior Instinct is concerned with finding the **right** way to do something, our Hunting Instinct looks for the **best** way. Those of us with Hunting Instinct take great pride in our abilities. We are motivated by need and supply, rather than loyalty, anger, or how we may look to others.

How far away does the woolly mammoth live? What kind of weapon do we need to kill it? How many people will we need to drag it back? What is the terrain like between here and there? Is there anything closer that we can hunt? What other sources of protein are easier to find?

Our Hunting Instinct values knowledge above all else. The more we know, the better we are at catching fish or snaring rabbits. What is the most efficient way to catch a fish? How can we catch the most fish with the least amount of effort? How can we breed fish to grow larger so we will have to catch less? How can we automate the whole process?

Those of us with Inventor Instinct can spend so much time thinking about these questions, we won't get any fish caught. Those of us with Warrior Instinct may brag about knowing the answers even if we don't, and those of us with Gathering Instinct will panic about catching enough fish. But those of us with Hunting Instinct will simply experiment and see what happens. In the end, we will bring home the fish. If we are nothing else, we are reliable and tenacious.

Because of this, our Hunting Instinct can make us great scientists. Attention to detail, careful calculation, and study are the perfect attitude for research. Our Hunting Instinct has a never-ending thirst for knowledge. We seek to become experts. We pride ourselves on proficiency and efficiency. We take most things very seriously, and have little humor. We value proof over faith, and feel satisfied when we figure something out, or perfect a technique.

Primitive

Primitive humans hunted for basic survival needs. The animals they hunted provided them with two things - food and clothing. By hunting large animals they could eat the meat and wear the skins. But, hunting also included smaller animals which are caught or trapped just for food - fish, birds, insects, worms, and larvae. Depending on where they lived, animals probably needed to be hunted on a daily basis in some tribes. Their Instinct would have lead them to find the best fishing holes, the best water holes where the large animals drank, and the best times to go hunting. Research, observation, tracking, and study of other animals' behavior is the specialty of the Hunter.

Primitive Hunters took great pride in their abilities, and decorated their weapons with images of their prey. After thousands of years of humans hunting on this earth, our behavior has not changed. Today, you can buy a hunting rifle with a laser-etched picture of a deer on the stock. We still decorate the weapons we use to hunt, with images of the animals we are hunting.

Modern

Most of us modern humans who live in civilized nations don't need to hunt for food. We now have industries involved in the production and distribution of food. Today, those of us who hunt for the meat and skins of other animal do so because we may have little or no money and can't afford to buy food. Or, we may live in a remote area, or simply want to supplement our diet with fresh caught animals.

We also now hunt for the sport of it. We experiment with different weapons, and practice tracking and stalking our prey. We research specific clothing and accessories to try to gain an edge on our target. We delight in a perfect hit, and feel pride bringing the animal back to show everyone else. Some of us big game hunters have the desire to kill the largest or most ferocious animal we can and mount it's head on our wall. We enter competitions, take pictures of the largest or heaviest animals we have caught, and spend hours recalling the stories.

But, today there is a different kind of hunting going on. The paycheck has replaced the animal for a lot of us. We don't have to hunt for meat, we can buy it. But we need money. The more money we have, the more meat we can buy. The more money we have the more of everything we can buy. We modern Hunters now hunt primarily for money. From investment bankers to bank robbers, our Hunting Instinct is the same; only the prey has changed. We use all the same skills of stalking, strategy, traps, lures, and sometimes physical force to bring home money. This is one area where our Hunting and our Warrior Instincts generate similar behavior.

The difference is that our Hunting Instinct justifies its behavior with facts and figures, and our Warrior Instinct justifies with attitude. In fact, our Hunting, Inventor, and Worker Instincts trust in facts and proof, whereas our Warrior, Gathering, and the Breeding

Instincts trust in myths and faith. It is a key part of how our Instincts work.

Those of us with Hunting Instinct can quote you statistics and numbers off the top of our head. We are the ones who can instantly tell you where the Dow Jones index was yesterday as compared to last week. We know what day the White Sale starts at Target and how much of a markdown there is going to be on towels. Numbers and facts are bread and butter to a Hunter. Survival for us depends on our ability to stay one step ahead of our prey, whatever our prey may be.

Typical Behaviors

This old house - Our Hunting Instinct is usually concerned with quality rather than quantity. What's important to us Hunters is that the job is done well. What is important to someone with Warrior Instinct is that the job is done. Those of us with Hunting Instinct will search the world for years looking for the best hand chisel for cutting blind dovetail joints. Those of us with

Warrior Instinct will run out on our lunch break and buy an air-powered nail gun.

The Habits of Highly Successful People - Really they can be all boiled down into one statement, "Follow your Hunting Instinct." Think about what you are doing: plan, research, and pay attention. It is rare for those of us with Hunting Instinct to not be successful. Our Warrior Instinct might make us perfect salesmen. But our Hunting Instinct makes us the perfect C.E.O. The key difference is vision. To work hard and be aggressive is absolutely necessary to be successful in business. But to have a clear vision of the future, and a strategic plan of action, is pure Hunting Instinct. This is why you don't want to play chess or other strategy games with us Hunters; you'll probably lose.

Geek Squad - Our Hunting Instinct likes things neat and tidy, well organized, evenly spaced, and concise. It likes clear goals, and proven methods of reaching them.

We are the tech heads. We understand how computers work, and can draw you a schematic of the flow of the data. Our minds operate like printed circuit boards.

We like order; we don't work well with chaos. We like to take things apart and figure out how they work. When we understand technology, it makes us happy.

Those of us with Hunting Instinct ask **how** something is done. Our Worker Instinct asks **if** something can be done, and our Inventor Instinct asks **why** something is done.

Our Gathering Instinct asks **when** something will be done, The three Breeding Instincts ask **what** can be done, and our Warrior Instinct asks **who** is going to do it?

Daddy Warbucks - Us Modern Hunters provide through our paychecks. Like our Primitive counterpart, the bigger it is, the more successful we will feel. Unlike those of us with Warrior Instinct who will boast and lie about having large sums of cash, we Hunters will actually have the cash. Our Warrior Instinct tends to be "penny wise and pound foolish"; that is, we obsess over trying to buy some things for a few pennies less, and meanwhile waste thousands on other purchases. Our Hunting Instinct, on the other hand, creates consistent bargain finders who can make budgets stretch like magic.

However, trying to bag the biggest prey we can takes time and commitment, which leaves little room for other things in a our life. Our Hunting Instinct, our like Inventor and Worker Instincts, focuses on the big picture. This can mean that we often miss the immediate issue at hand, which might be the happiness of our spouse and family. We may pride ourselves on being good providers, and lose the ones we are trying to provide for, by trying too hard to provide for them.

Flashing a Wad of Cash – Houses and cars can be bought on credit. Those of us with Warrior Instinct can pump up our appearance by surrounding ourselves with things we don't actually own, but are making payments on. Those of us with Hunting Instinct, on the other hand, may not feel secure if we didn't actually own the house and car. We would rather save and then buy things with cash. We may not trust credit at all and instead carry around large amounts of money. We Hunters feel good about ourselves when we actually have the money to afford to buy things right out. We may also hang onto money believing that even though we have plenty now, lean times may be just around the corner. We will feel satisfaction from still having plenty when others are broke. This can also make us feel like a good provider to our family.

Clearance Sales – Those of us with Hunting Instinct are also Hunters of bargains. If there is a coat that usually sells for $200 and is marked down to $125, we may buy it and feel good about the $75 we saved, not the $125 we spent. Even though we had not planned on spending $125 and we will have to make up for it somewhere, and even though the $75 doesn't actually exist, we are still motivated out of our Hunting Instinct. We Hunters take great delight in saving pennies. Shopping discount houses, buying in bulk, and buying generic brands are all be behaviors motivated by our Hunting Instinct. This is one area where our Hunting and our Gathering Instincts look alike, but our personal payoff is different. Those of us with Gathering Instinct are motivated to have enough stuff on hand so we don't run out. Our personal payoff is knowing we don't have to worry about gathering for a while. For those of us with Hunting Instinct, our personal satisfaction comes from how efficiently we pulled it off.

Stuffed and mounted – For some of us men, having the best looking wife in our circle of friends is a sign of how good a Hunter we are. We men sometimes use the phrase "Trophy Wife" in describing the wives of men we know. Likewise, for some women, having a successful attractive husband is a sign to their friends of how good a Hunter she is. Some women actually talk about what a great "catch" he is

Work hard, play hard – Our Hunting Instinct makes us better athletes than our Warrior Instinct. Where one of us with Warrior Instinct might go bungee jumping off a bridge to impress our friends and prove to them that we're not afraid, one of us with Hunting Instinct would do it just to test our own limits. The behaviors are the same, but remember, these Instincts are about motivation. Our Warrior Instinct is concerned about competing with others, and looking good in front of other people. Our Hunting Instinct is concerned with personal testing. The skinny guys from Kenya who win the Boston Marathon probably have Hunting Instinct. Marathons are about endurance, and those of us with Hunting Instinct generally have more endurance than anyone. It is our nature. Us big beefy guys with Warrior Instinct who can bench press our own body weight don't win marathons. Most of us drop out before the finish, or slug it out for four or more hours just to prove to everyone watching that we could do it. Us skinny winners would run the marathon, even if there was nobody watching. What's important to us is that we improved on our usual time, and we do it by calculating ways to shave a few more seconds off each mile. Efficiency, efficiency, efficiency.

Not asking for directions – As primitive people, we wandered the earth always in search of food. For the Hunter, this was their job. If we cannot provide, we lose our sense of self worth and importance to the tribe. For 3.5 million years, there was no one to ask directions from. Remember, Instinct usually responds before common sense. Not to mention, if we find a good fishing hole we probably don't want to tell anyone else how to get there. Our Warrior Instinct, on the other hand, won't ask for directions because we think we are supposed to know the right way to go instinctually. For us, it's about saving face.

Not turning on lights when entering a room – Primitive wild dogs used to sleep in tall grass to hide from predators. They would turn around in the grass before laying down to flatten a space. Domesticated dogs still turn around before they lay down. Even though they don't sleep in the grass any more, it is an instinctual leftover from the past. We are more like the other animals than we are different. Primitive Hunters would not have gone hunting mammoths with a flaming torch in their hand so they could see where they were going. The light would have either warned off their prey, or caused it to attack them. And so, like domesticated dogs, us Modern domesticated humans with Hunting Instincts may not instinctively turn on a light. Likewise, those of us with Warrior Instinct may not want to draw attention to ourselves either. We don't want the enemy to see us coming. Remember, we have been on this earth for almost 4 million years and Thomas Edison didn't invent the electric lamp until 1879. Old habits die hard.

Compatibility

Never enough meat - Those of us with Hunting Instincts work ourselves to the bone thinking what a great provider we are being to our family. But this means that we may spend less and less time at home. Our relationships could start to suffer as the intimacy level drops. If you confront us about this, we may be confused and angry - "I'm working my butt off for this family and all you can do is complain!" We may truly see that what we are doing is what we are supposed to be doing, and you should be grateful for it. The flip side to this is the thing which will bind you to us. We are the best and most reliable providers on the planet. And we will actually work ourselves to death to meet your needs.

Partners

Hunting and Hunting – As a research team, hard to beat. As a passionate couple, we could spontaneously combust. As long-term partners, we could be famous. We would probably be the kind of couple that others see as anti-social. We would need very little outside influence in our relationship. It would be continually stimulating for both of us.

Those of us with Hunting Instinct are reliable, committed, and supportive partners. If the relationship works for us, we will never leave it. It is an internal directive motivated by our Instinct. We pride ourselves on doing things well, and making situations work the best we can.

We can be fascinating dinner guests, if you can get us to talk about the things we have studied and researched. However, if you play against us as a team on board games or trivia games; you will most likely lose. Our children can be insufferable book worms and social outcasts, or rebellious dropouts who feel like they can never measure up, so why even try.

Hunting and Gathering – Those of us with Gathering Instinct can be highly attracted to someone with Hunting Instinct because of their dedication and relentless pursuit of what they want. But, the difference in our perspectives could be a stumbling block. Hunting Instinct focuses on the big picture, and Gathering focuses on the issue at hand. Those of us with Hunting Instinct may also feel disappointed in our Gathering partner's blind acceptance of unproved theories. Those of us with Gathering Instinct accept things on faith. Our Hunting Instinct questions everything before we will accept it. But there are other stumbling blocks. Those of us with Gathering Instinct may display a lack of interest in furthering our self-education. This can be disappointing to our Hunting Instinct partner who values knowledge above all else. Furthermore, those of us with Gathering Instinct can have a preoccupation with stuff, and live with a certain level of clutter in our lives, but Hunters tend to be minimalists and travel light so they can move efficiently. Could cause problems.

Hunting and Warrior - These two can have problems with who is in charge of what. Those of us with Warrior Instinct may not be as efficiency-minded as someone with Hunting Instinct, and may take offense if our partner makes us feel like we are doing something wrong. While those of us with Hunting Instinct may be clever and use strategy to get our needs met, we might often bump into the black and white thinking of Warrior Instinct over many issues. Warrior Instinct has difficulty with long term relationships. If these people start to believe their partner is against them, the intimacy level can disintegrate. Someone with Hunting Instinct, on the other hand, will tend to be tenacious and dedicated. If they strongly believe the relationship is what they want, they will do whatever it takes to make it work - short of being abused, that is.

Hunting and Worker – Both of us share a wide perspective on reality. However, those of us with Worker Instinct tend to believe that other people control the details of our lives and our destiny. Meanwhile, those of us with Hunting Instinct believe that we are the captains of our own ships, and chart our own course. Those of us with Worker Instinct tend to let things slide and not worry about stuff too much, because we can't change it anyway. So, we are basically easy going and easy to be with. We may enjoy our Hunting partner's ability to find a bargain, because that can be one small way we get to "win one" when we are so used to losing. Those of us with Hunting Instinct, however, may see some of our Worker mate's behaviors as lazy or inefficient, and it may cause conflict.

Hunting and Inventor – This could be an excellent match. Both of us would have a broad view and could work well as a team. If we play on each other's strengths and don't get jealous, it could work quite well. Some of the best research teams in the history of inventions and discoveries have come from this combination. One person has the vision, and the other has the determination to do the research to prove it. If this passion for work also folds over to an emotional passion for each other, this can be a very rewarding and exciting relationship. One major problem area is the difference between knowledge and wisdom. Those of us with Hunting Instinct prize knowledge as the secret to life, whereas those of us with Inventor Instinct prize wisdom over knowledge. Since these are central parts of our respective instinctual behaviors, they may continually cause us conflict, and could kill the intimacy level.

Hunting and Attraction - Powerful initial attraction for both. The self-confidence and commitment of us Hunters will be comforting to our mate. Those of us with Hunting Instinct will most likely assume the dominant position in the pair regardless of our sex, and we can operate very efficiently and smoothly as a couple, with a clear and equitable division of duties. But we can become very jealous and uncomfortable watching other people check out our attractive mate. Our Attraction Instinct can cause jealousy which could bust us up.

Hunting and Mating - This would be a good setup for those of us with Hunting Instinct because it's an efficient way to get our sexual desires met. And it's good for our partner because we Hunters are not notoriously mushy or demanding. The only booger is loyalty. The one with Mating Instinct cannot fool around with someone else - Those of us with Hunting Instinct don't like betrayal. We are likely to walk out without warning, but then retaliate in strategically destructive ways. Hunting Instinct is masterful at revenge. This can be a perfect couple to have a long-term affair if they are both married to other partners.

Hunting and Nurturing - See Nurturing and Hunting - page 207.

Hunting Instinct

In a Nutshell:

Hunting Instinct focuses on
efficiency, details, evidence, and proof.
It desires information, knowledge,
and order.

Hunting Instinct is satisfied when:

We master a technology.
We improve the efficiency
of an activity or process.
We find a bargain.

When it's needs aren't met:

We use strategy and manipulation.

Given unlimited resources:

We could become so educated
that we would not be able
to communicate with each other.

Gathering Instinct

Gathering Instinct is our support system. We gather every day in order to have something to eat and drink, and a place to sleep. But, our Gathering Instinct can never be satisfied, because no matter how much stuff we gather, sooner or later we will need to gather more. We cannot possibly stock up enough food to last us for the rest of our lives. Thoughts like this can make those of us with Gathering Instinct believe that there will never be enough of anything. Consequently, our Gathering Instinct is the basis for all kinds of anxiety and panic. We gather out of fear of running out, and so we often gather beyond what we actually need. Because of mass production, we can now gather what we desire as well. And some of us simply gather out of sheer force of habit. This is an Instinct that operates without reason or rationalization. Our Gathering Instinct makes us buy something simply because it is cheap. Unlike our Hunting Instinct, which can spur us to shop for the bargain when we need something, our Gathering Instinct has a narrow and short-sighted view. Those of us with Gathering Instinct will buy things we don't need simply because they are on sale. Our houses are usually full of things that, "Somebody might need someday." Our Gathering Instinct delights in having more than enough.

We fill our lives with stuff, and spend a considerable amount of time moving stuff from one place to another. We can often be late, because the stuff we need to organize before we do anything can be overwhelming. Those of us with Gathering Instinct are like squirrels, dashing back and forth with our cheeks bulging with acorns.

PRIMITIVE

We used to gather water, firewood, nuts, berries, building materials, and pretty much whatever was needed at any given time. Since everything we gathered got used, a never-ending supply was needed. Like every other species on the earth, every day we had to eat and find somewhere safe to sleep. This was a never-ending task. There was no manufacturing and distribution, no grocery stores, no housing, and no transportation. So most of our day was spent gathering something. And then we would wake up the next day and gather more.

MODERN

Gathering today is a lot simpler than it used to be. We invented mass production and distribution. Now we have stores that stock and sell all kinds of things. Food is everywhere, and other people manufacture our clothes and furniture for us. All we have to do is pay for it and take it home! Life is great! But still, we modern humans with Gathering Instinct can have canned goods stocked up in the cellar, boxes of old clothes in the attic, and money stashed in the mattress. Even though we have access to whatever we need in most western civilized countries, our Gathering Instinct is still not satisfied, and inspires us to stock up and hoard essentials. It is knee-jerk instinctual behavior, and only focuses on the immediate issue at hand. Our behavior can run the spectrum from a Boy Scout mentality of being prepared for anything to a compulsive pack rat who just can't seem to throw anything away. Our Gathering Instinct prompts us to buy on impulse. We live for a sale. Two-for-one, 10 percent off, clearance, and mark downs are the language of joy to our ears. "If I get all my Christmas wrapping paper on sale the day after Christmas, then I don't have to worry about it next year!"

Those of us with Gathering Instinct might be attracted to a job in purchasing or inventory control. Making sure a store has enough product on hand at all times would be second nature to us. We are the Office Manager that is the backbone of the whole company: "I don't know how we would operate without her." Our

Gathering Instinct drives us to do whatever we can to take care of things so we don't have to worry. The only problem is, because it is our nature, there is usually something new to worry about just around the corner.

Our Gathering Instinct is also the root of anxiety, fear, and panic. It may well be the cause of some of our emotional and psychological conditions such as panic attacks, chronic anxiety, and obsessive compulsive disorder.

Typical Behaviors

All-You-Can-Eat Buffet - For millions of years it was a daily job to go out and gather whatever we could find to eat. And when we did find food, we ate as much as we could, because we never knew where or when our next meal was going to come from. This is the Instinct which drives the buffet line.

This is not about rational thought or common sense. Common sense would tell us that if we weigh 300 pounds we probably don't need to keep eating all we can eat until we feel full. But even those of us that are not "overweight" flock to the buffet line. If we have Gathering Instinct, this is heaven for us. All the food we could possibly ever want, and all we have to do is eat it. And someone else will even wash the dishes! After 3.5 million years of scrounging around in the bushes trying to find enough berries to quiet the rumbling in our stomachs, finally there is enough food for everyone! Or is there?

I JUST CAN'T LOSE THE WEIGHT -

Of course we can't! Our Gathering Instinct is in charge. Many of us have chronic problems with our weight. Our Gathering Instinct tells us that there is never enough, grab all we can now, because there may not be any later. Because, for 3.5 million years there probably wasn't enough. This is about survival. Just because we now have fast food restaurants right around every corner doesn't mean the Gathering Instinct will disappear. Just because we have hospitals and schools doesn't mean our Nurturing Instinct will disappear.

Eating all we can when it's available was probably not a problem when we were eating fresh salmon from a stream, and handfuls of berries and nuts. But we have changed what we eat. We have developed food that satisfies our Gathering Instinct more effectively, and in doing so have started to eat things not normally found on the earth. Instead of eating whole grains, we have developed refined grains which are easier to make bread out of. We discovered we have a desire for sweet stuff, and figured out how to mass produce and refine sugar. White flour and white sugar have become staples in our modern Gathering diet. We have modified and refined natural foods in order to mass produce something to satisfy our Instinct. We want food available everywhere and at any time. And we want food which feels filling in our stomachs. The goal is to never ever feel hungry again.

South American Indians first cultivated potatoes. There is no record of potatoes beyond about 4500 years ago, and white people of European origin have only been eating potatoes since about 1500 AD. The point is, our bodies aren't used to this stuff, because, for 3.5 million years, we never ate it. The same with processed white rice and cow's milk. White rice is faster to cook so it can satisfy us quicker. Milk-producing cows are the result of cross breeding, and didn't exist for the first 3.5 million years. And how many meals a day do we eat that contain white flour, sugar, cow's milk, potatoes, or white rice?

Of course we are gaining weight; our Gathering Instinct is still driving our behavior, but we are eating different foods. Meals made with potatoes and bread make us feel full and happy. They please our Gathering Instinct, and so we eat them. And diets don't work because our Gathering Instinct will not be satisfied with carrot sticks and celery! They simply do not feel the same inside our stomachs. And diets based on regulating the **amounts** of white flour and potatoes we eat don't work, because our Gathering Instinct doesn't feel satisfied unless we feel full. We cannot short circuit our Gathering Instinct with reason and rationalization. Our brain is not in charge, our Instinct is. Our Gathering Instinct is only satisfied when our stomach feels full. If we even start to feel slightly hungry, or more specifically, **less full**, our Gathering Instinct will motivate us to eat something. For some of us with Gathering Instinct, we need to feel full all day. If we start to feel the least bit hungry we will eat something. If we are constantly trying to maintain a non-hungry feeling, we will gain weight and not be able to lose it.

The Only Diet That Really Works!

Don't eat anything that has not been growing on this planet for at least as long as we have.

Why This Diet Will Not Work

Our Gathering Instinct won't like it.

We are more like the other animals on this planet than we are different. We are like dogs that have developed a taste for people food. But, unlike dogs, we have developed ways to produce and distribute this food to thousands of convenient locations so we never have to worry about being without food again. We eat until we feel full. Bread, potatoes, and rice make us feel full. They feel warm and comforting inside our stomachs. They may not be nutritious, but it doesn't matter to our Gathering Instinct. What matters to our Gathering Instinct is feeling full, and that is why we gain weight and can't lose it.

And, if we also have Nurturing Instinct, we put ourselves through a regular cycle of shame, guilt, and fear over our weight. We shame ourselves about our inability to lose weight, guilt ourselves when we eat things that aren't healthy for us, and live in fear of the medical consequences of being overweight. This feels normal to those of us with Nurturing Instinct. But, it will not help those of us with Gathering Instinct to lose weight or stay on a healthy diet; it will simply create a cycle in our lives of losing and gaining, and feeling shame, guilt, and fear.

And, if we have Warrior Instinct as well, we create more problems for ourselves. Force, or threat of force, is how our Warrior Instinct operates. Our Warrior Instinct will tell us that we can't lose weight or stay on a diet because we aren't trying hard enough. It believes that we can force ourselves to lose weight, or stay with a diet and exercise program. The idea that we don't have enough "will power" to avoid overeating is our way of explaining that our Warrior Instinct has no power over our Gathering Instinct. Force works for those of us with Warrior Instinct. But force doesn't work for those of us with Gathering Instinct, because Gathering is instinctually driven behavior, it is not a matter of conscious thought or reason. And besides, for over 3 million years we have never had to concern ourselves with exercise or trying to not eat processed foods.

We are the architects of our own misery. When we created civilization and the concept of working for a living, we also dramatically changed when we eat and how we eat. We fit our eating in and around our working. We need something fast and filling for breakfast because we are going to be late for work. We need something fast at lunch because we only get a half hour to eat, and we really should work through lunch to get that report finished. We need something filling for dinner because we have rushed all day and missed lunch and we're starving. So we fill ourselves up and then lay down and go to sleep. Our desire to make money is dictating our nutrition, because it is dictating the hour-by-hour structure of our day. Our Warrior, Gathering, and Nurturing Instincts are the source of our behaviors around eating. And we can see the consequences every time we look in the mirror.

THE MYTHICAL CAVEMAN - We can have a nostalgic view about our ancient ancestors. We may believe they were all lean and fit and healthy, and somewhere along the line we invented sofas and television and became lazy and fat. The truth is, it's just as likely that for 3.5 million years we would catch an animal and cook it, and then lie around in our cave stuffing ourselves and sleeping. Remember, we had no money, no where to work to make money, no need for money because we couldn't buy anything anyway, and we were dead at 25. How many overweight people do you know under the age of 25? What if the percentage of overweight people under the age of 25 is still the same as it ever was? Did cavemen work out? Did they diet? Did they shame themselves about being "overweight"? Were we happier and healthier? Or sadder and sicker? Or pretty much the same?

FAMILY SIZE - Buying in bulk is the next best thing to eating in bulk. To those of us with Gathering Instinct, a hundred rolls of toilet paper means not worrying about having to buy it for a long time. And that's a good thing, because our Gathering Instinct will worry about almost everything. Two gallons of ketchup will probably spoil before it all gets used, but it will satisfy us to know that we don't have to worry about buying ketchup for a long time.

Rush Delivery - Another part of the Gathering Instinct is timing: We have to have it right now! Today we are hungry, food coming tomorrow is no good. What are we going to eat today? We need firewood and water right now! This is the root of immediate gratification. Convenience foods, convenience stores, rush delivery, fast food, and credit cards, all feed this desire to have it right now! There is nothing logical about paying more money to have something sooner. This behavior has grown out of the primitive notion that if we don't grab that bunch of berries right now, they won't be there tomorrow, because a bear might eat them.

Gambling - While this may seem a risky thing to do, it's a question of motivation. Those of us with Worker Instinct would probably be too cautious to gamble, but those of us with Gathering Instinct will take any short cut if there is one available. Anything that can be done to take the pressure off of having to gather more is a good thing. If we can buy a week's worth of groceries with our paycheck, that's great. But if we gamble $20 and win enough money to buy a year's worth of groceries, all the better! The carrot dangling at the end of the stick is highly attractive to our Gathering Instinct. Pull tabs, scratch cards, casinos, lotteries, and frequent shopper cards all play on this Instinct. It is the payoff that is our motivation. If we can gather more stuff with less effort then sign us up!

VIVA LAS VEGAS !

Las Vegas is a major mecca for our Gathering Instinct. Here we can run wild in a place which is specifically designed to encourage our Instinct. The plane fares are dirt cheap, the hotel rooms are cheap, and the drinks are free if we are gambling. Free is a word that we Gatherers love to hear. But, none of it is really free.

The hotels in Las Vegas are also casinos, restaurants, and bars all in one. If we stay in the hotel, gamble in the casino, eat in the restaurant and drink at the bar, they are taking a great deal of money from us. If our hotel room only costs $40 a night but we gamble $250 downstairs at their tables, then we have spent $290 in the hotel. Our Gathering Instinct will typically not see money that has been gambled as money that has been spent. In our mind, we see it as somehow different money. But, if we are sticking our hand in our pocket and pulling out money and handing it over for one thing or another, does it really make a difference where it goes? The point is - **we** no longer have it.

A typical weekend at a Las Vegas hotel can run us $100 for air fare, $80 for the room for two nights, perhaps another $100 in food and drinks, and $30 in tips.

Grand total - $310. If we don't gamble!

For $310 we will have to search pretty hard for a good deal on a weekend in a nice hotel in any other city of the world. But we will probably not sit in their lobby all weekend dumping $1000 into a machine. Let's say we gamble an average of $500 a day. So our $310 weekend has actually cost us $1310, or maybe as much as $1800 if we arrive on Friday with plenty of time to gamble before dinner. So our room and food and drinks, even if they are free, still cost us $500-$750 a night!

At those rates, the hotels in Las Vegas can give away rooms for free, feed us for free, and fly us in for free, and still make a good profit. So what do they care if they slide us a free beer while we are playing poker. But all our Gathering Instinct sees is, "Hey, I got a free Bud Light!"

Bulls and Bears - The riskiest form of gambling is the stock market. If we win, we can win big. But if we lose, we can also lose big. This is not putting in $20 and winning a million. This is putting in our life savings and watching it dwindle down to nothing, or double or triple or quadruple. But to our Gathering Instinct, it can be the same motivation. If we can invest part of our salary and double or triple it, we will probably give it a try. We will typically look at the winning side and rarely the losing side. And just like other forms of gambling, it's always the next one that's going to hit big and all our problems will be solved. Those of us with Gathering Instinct will usually have a great deal to say about what we are going to do with all the money we make when the strike it rich, but very few of us will have a plan B standing by for when we lose our life savings.

WE CAN NEVER BE TOO THIN OR TOO RICH - If the "all-you-can-eat" buffet satisfies our Gathering Instinct, then this could lead to compulsive overeating. Overeaters Anonymous meetings are full of those of us who simply cannot stop eating for one reason or another. Some of us overeat to make ourselves unattractive to the opposite sex because we were molested as children. Some of us overeat because we grew up in families that overate. Some of us substitute food for other things: "If I can't feel full of love, then at least I can feel full of food, and that makes me happy." Some of us can control our behavior with help, but some of us cannot. Some of us overeat simply because we have no mechanism which tells us how much is enough. Indeed, the whole concept of the Twelve Step program is to admit that we are powerless over our behavior. Perhaps what some of us are really powerless over is our Gathering Instinct.

And, on the flip side, if there is no understanding of how **much** is enough, there may also be no understanding of how **little** is enough. Anorexics and Bulimics also talk about feeling powerless to control their behavior. If we have Gathering Instinct and also Attraction Instinct, it could create opposite behaviors inside of us. This could explain the binge/purge cycle that some of us wrestle with. Gathering Instinct binges and Attraction Instinct purges. Remember, this is instinctual behavior. It is not based in logic or reason. We are powerless over it. We could be trying to live with two separate conflicting Instincts.

Our Gathering Instinct could also drive those of us who compulsively overspend. Debtors Anonymous meetings are full of those of us who have no cutoff mechanism which tells us when we have enough. We live from day to day and buy and spend like there is no tomorrow. Our focus is on the here and now, and not on the bigger picture. Again, some of us overspend to compensate for a lack somewhere else in our lives. But there could be some of us who are simply acting out of a runaway Instinct. Some of us in Debtor's Anonymous also attend Overeaters Anonymous and other 12 step programs. We may be separately treating the behaviors associated with a common Instinct: Gathering. No such thing as enough, can mean no such thing as enough of **anything**.

THE AMERICAN DREAM - Workaholism could also be fueled by our Gathering Instinct. If I we make $60,000 a year, then we can work a little bit harder and make $70,000 or $80,000, or $90,000, and get a bigger house, another car, and a newer set of golf clubs. We may even say that we are driven to succeed. What we may actually be driven towards is simply Gathering more, and more, and more. Those of us with Warrior Instinct can become workaholics because we like the power that money gives us. We are usually motivated out of the desire to look good in front of other people. Those of us with Gathering Instinct, however, are generally motivated out of the belief that there is no such thing as enough. Once we become successful, do we stop? There are many of us who are successful and have large sums of money in the bank. But we don't rest. We are still working, investing, starting newer and larger projects, building more expensive houses so we have to keep working to pay for them. Many of us are so mesmerized by our Gathering Instinct that we keep ourselves poised just one paycheck away from losing it all, and ending up homeless. This kind of behavior is rewarded in America, because it supports the "American Dream" that if we work hard enough we can get whatever we want. This could be the green light for our Gathering Instinct to run wild.

COMPATIBILITY

THE BOY SCOUT - " Always Be Prepared ", is a motto that was probably invented by someone with Gathering Instinct. Those of us with Gathering Instinct can be a joy to live with or a pain in the ass. If you appreciate having a hundred rolls of toilet paper in the linen closet then it's a good match. But, we can also be the one who wants to plow up the front lawn and plant tomatoes.

THE PACK RAT - We may have boxes of old coats stashed in the attic and drawers crammed with silverware that doesn't match. We never throw anything away because we may need it tomorrow. We shop at flea markets and thrift stores, and fill our houses and lives with stuff. Those of us with Gathering Instinct are like goldfish; we can grow as large as our bowl will allow: Amelda Marcos had 10,000 pairs of shoes.

THE WORRYWART - Gathering Instinct can make us become obsessive. The prospect of having to gather on a daily basis can wear us out. Since our job is never done, there is usually something new to worry about all the time. We are the ones who lay in bed at night thinking, "Do we have enough milk for breakfast? Did I remember to mail those bills today? What am I going to get my sister for Christmas? Is California going to fall into the ocean? And if it does, where are we going to get avocados from? What is this small red mark on my arm? Am I getting cancer? Maybe I'll just get up and check the lock on that back door again."

COMMITMENT - All this having been said, those of us with Gathering Instinct can be a very stable partner. We can be tenacious and will probably do whatever it takes to keep the relationship going. If you can live with a few quirks, you can have a long satisfying life, and never have to worry about running out of toilet paper.

PARTNERS

GATHERING AND HUNTING - See Hunting and Gathering - page 30.

Gathering and Gathering - No room to sit in our car. A path that winds between piles of stuff from the front door of the house to the kitchen. Every horizontal surface stacked with magazines received but never read, unfinished projects, newspapers, photos, old mail, catalogs, and books. A garage so full the car has to stay outside. A basement that the Fire Marshall would have a heart attack over. An attic the same. Kitchen closets full of groceries, some of them years old and never touched. More clothes on the floor of the bedroom than in the closet. No room to set a glass down on the night stand. Stuff, stuff, stuff. Under the bed, behind every door, on top of the fridge, next to the toilet, on top of the TV, and slid down behind the radiator. And two of us, happy as clams, carrying three bags full of stuff with us wherever we go. Hey, if it works for you ...

Gathering and Warrior - Both of these Instincts see the immediate issue and are preoccupied with their own concerns. Those of us with Warrior Instinct tend to see people with Worker and Gathering Instincts as someone to serve us. It is highly unlikely we would pair romantically, unless the one with Gathering Instinct has physical beauty and can boost the way other people look at the one with Warrior Instinct. At best, it will be a shallow relationship. But, someone with Gathering Instinct may be attracted to the self confidence and take charge attitude of the one with Warrior Instinct. They can see this as reassurance that they may never want for anything again. Unfortunately, it could turn into a moth-to-the-flame relationship.

Gathering and Worker - See Worker and Gathering - page 118.

Gathering and Inventor - These are partners with opposite views (broad and narrow), and they may not be able to weather the constant perspective difference between them, unless each one has another balancing Instinct. Our Inventor Instinct moves quickly from one thing to the next, needing a constant input of new things to consider. These people like theories and intellectual stimulation. The one with Gathering Instinct may find it hard to keep up with them, and can feel lacking in their ability to keep their partner interested. Also, Inventor Instinct tends to motivate people to keep their lives as simple and low maintenance as possible, whereas Gathering Instinct desires more and more. Their levels of maintenance are dramatically different. Could become an issue.

Gathering and Attraction - Those of us with Attraction Instinct are focused on how we look. Those of us with Gathering Instinct rarely even consider how we look. No common ground at all, not even a workable opposite to balance each other out. If this pair ever did hook up for some reason, the one with Attraction Instinct could start to pressure their partner, which might drive them into compulsive behavior.

Gathering and Mating - A good steady partner for the one with Mating Instinct, but a potentially damaging arrangement for someone with Gathering

Instinct. Those of us with Gathering Instinct can be drawn to having sex in order to fill an empty feeling we have inside. Those of us with Mating Instinct will gladly fill that hole for them. This is the kind of relationship where those of us with Gathering Instinct might feel unloved, possibly because we are overweight, and we like the attention from the one with Mating Instinct because it makes us feel lovable. But our Mating partner may only want to have sex with us, and not want to be seen in public with us. And so we are emotionally torn. We may want to pressure our sex partner to acknowledge our relationship in public, but we are scared that if we do, we might lose them. And we don't want to lose the sex because it fills us up and makes us happy, even if it is only for short periods of time. The double-edged sword. This can easily become a relationship where those of us with Gathering Instinct gain a bunch of weight because we don't feel loved. Or we may turn to alcohol, drugs, gambling, or spending money. If our Mating partner doesn't want to mate with us, we can feel that old familiar feeling like there's never enough love.

GATHERING AND NURTURING - See Nurturing and Gathering - page 207.

Gathering Instinct

In a Nutshell:

Our Gathering Instinct makes us believe
there is no such thing as enough, or too much.
It motivates us to always have a full stomach,
and a house full of stuff.
It makes us live with a constant level of
anxiety, and fear of running out of things.

Gathering Instinct is Satisfied When:

We aren't hungry.
We have a full refrigerator.
We don't feel like we need anything.

When It Isn't Satisfied:

We use begging and pleading.

Given Unlimited Resources:

We will stuff ourselves
until we are too large to move.

Chapter Three

Warrior Instinct

WARRIOR INSTINCT IS CONCERNED WITH SAFETY AND SECURITY. Those of us with this Instinct are motivated to protect the things we feel are our possessions. This can include other people, as well as objects and natural resources. Our Warrior Instinct is territorial. It makes us see people as either friend or foe, with us or against us, and there is no middle ground. We can see the world as good or evil, wrong or right, black or white. There is no gray area. This is exactly what makes those of us with Warrior Instinct effective at protecting people and resources. Warrior Instinct is indispensable in society. It spurs us to stop people from other tribes from stealing our stuff, abducting our partners, and wiping us out.

Our Warrior Instinct is the origin of our feelings of loyalty, dedication, and personal boundaries. It can come out when we feel cheated, used, compromised, ignored, or abused. Our Warrior Instinct is the Instinct which motivates us to fight for survival. It is also the Instinct which motivates us to be heroes. It is the thing which makes us jump in a river, and save a drowning person. It makes us put personal safety and consequences aside, and spurs us to action, because it's the right thing to do.

Those of us with Warrior Instinct will tell you we are concerned about your security. But, concern for your security is actually motivated out of our own feelings of *insecurity*. Our Warrior Instinct can make us feel constantly insecure. It is how some of us stay sharp and alert. If we are police officers, our Warrior Instinct can prompt us to carry guns to protect ourselves from criminals with Warrior Instinct. We use handcuffs and restraints to protect ourselves from our prisoners. We lock people in cages, so we will feel safer walking the streets. And we soldiers with Warrior Instinct simply kill the people we feel threatened by.

Those of us with Warrior Instinct are motivated out of fear. For instance, we would deny, cover up, and destroy evidence of extra-terrestrial and other unexplained phenomena. If we cannot explain it or control it, it scares us. Anything that scares us is seen as an enemy, and it must be destroyed. This is the black and white thinking of Warrior Instinct. It is instinctual, not logical. Feelings of not being safe, and not trusting other people, are the prime motivators to creating safety and security. It is ironic that the most insecure people on earth are the ones we put in charge of our security. And it is a never ending task, because our Instinct can not be satisfied. So, those of us with Warrior Instinct are always vigilant.

Our Warrior Instinct is often judged by our Nurturing Instinct as being bad and causing problems. This disagreement between Warrior and Nurturing Instincts over how to handle situations is as old as human beings. Some have called it the battle of the sexes. It is actually the battle of the Instincts, or more specifically, their perceptions. Each of us believes we are right, and will argue the other to death. This conflict is compounded by the seemingly magnetic attraction that those of us with Nurturing Instinct have towards those with Warrior Instinct. Women with Nurturing Instinct are still physically attracted to the big strong muscular Warrior who can protect them and their children. As the natural biological partners to produce the healthiest children, this attraction has served our species well in primitive times. Beyond the age of 25, however, it seems to cause us more problems than it fixes.

The fight between our Warrior and our Nurturing Instincts continues to rage over any issue from squeezing toothpaste to foreign policy. Rising to defend what we believe is an instinctual behavior. You can try to shame us out of our Instincts. You can imprison us. You can send us to therapy to learn different ways to deal with our anger. But even a calm, sane, rational 95 pound woman will pull the trigger if someone is trying to kill her child. This is an Instinct, pure and simple. It is alive and well, and has kept us alive for a very long time.

PRIMITIVE

A Primitive Warrior would have served essentially one purpose - to protect the other members of the tribe. This person would not have been afraid to kill, and they could have been killed themselves at any time, so they lived with a constant level of fear and vigilance.

The Warrior may have been the tribal leader, due to their fearless nature and willingness to do whatever it takes to protect the others. A strong Warrior would be someone to rally around and believe in. This belief, and the Warrior's willingness to live a life of violence, are part of the reason we have survived for millions of years.

The other thing that Warrior Instinct does is thin us out. Apart from Warriors from other countries, viruses, and a few large animals, we have very few natural predators. As a species, human beings police their own numbers. Through warfare, crime, and passionate rage, a random segment of the population is regularly killed. Now we live much longer than we ever have before; Nurturing Instinct has made great strides in extending human life. So, the Warrior Instinct is unable to keep up with thinning us out, and our population is over six billion.

MODERN

We modern humans with Warrior Instinct seem to have changed very little from our Primitive counterparts. Being ready to fight anyone at any time is instinctual behavior. We cannot intellectualize it. It is all relative to the individual and what we believe. If this was not true, there would be a lot less people in prison.

If you train us to kill without hesitation; we will. We will do it on the battlefield, and we will do it in the back yard. Even though physical violence is the least effective means of conflict resolution, it still continues to be used. Warrior Instinct makes us believe that all we need to do is shoot a few more bullets, or drop a few more bombs, and then it will all be over. But wars don't seem to end the need for more wars, because our Warrior Instinct doesn't go away - the need to feel safe, is never satisfied. Never. There is always a new enemy just around the corner.

Our Warrior Instinct will never be satisfied in the same way that our Nurturing Instinct can never be satisfied. It is inconceivable to our Warrior Instinct, that there would ever come a time when there would be no more enemies. In our mind there is always a need for weapons and defense. It is inconceivable to our Nurturing Instinct, that there would ever come a time when we wouldn't need health care. In our mind there will always be one more thing we can do to make people live longer and healthier lives. Both are instinctually driven, and thus neither can be quenched.

Our Warrior Instinct manifests itself in different ways since we have become civilized. The basic key to understanding Warrior Instinct is this:

Those of us with Warrior Instinct
look at the world and see enemies.
It is only the nature of the enemy that differs:

- The guy who stole our parking spot.
- The woman who stole our husband.
- The other football team.
- Illiteracy.
- The Devil.
- Fat (or now it's carbohydrates)
- A disease (The Battle against Cancer).
- A concept (Lawyers who fight for justice).
- A way of life (The Fight Against Poverty).
- Irresponsibility (Mothers Against Drunk Driving).
- A nationality (The Serbians).
- A race (Blacks).
- A religion (Jews).

And so on.

Typical Behaviors

Every Mistake In The Book - Have you noticed that this chapter looks a little different than the rest of the book? Evidently, there are some people in the book publishing business with Warrior Instinct who believe that there is a "right" way and a "wrong" way to put a book together. And apparently I have done it "wrong". So, I have tried to layout this section on Warrior Instinct the "right" way. Here are some Warrior Instinct rules for book publishing:

Justifying text - Notice how the words in this section line up magically on both the left *and* right side of the page. Evidently, this is the professional way to make a book. The words are lined up this way by the software used to layout a book. The way the software lines the words up is by stretching out or scrunching up the space between the words. For comparison, look at the previous section on Gathering Instinct and notice the jagged edge of words on the right hand side of the pages. This is considered "wrong" because it doesn't look as good. Personally, I find it more distracting to see words randomly scrunched together and stretched out. I prefer to see words evenly spaced on a page. I also find straight edges on both sides of the page to be dull and monotonous. Furthermore, if you are paying that much attention to the *appearance* of the words, are you missing the *content*?

Italics not bold - Apparently it's "right" to use *italics* for emphasis, and "wrong" to use **bold**. I just don't like the way italics look. Plain and simple. I'd rather use **bold**. I think italics are more difficult to read. They look weak and not emphatic enough. And anyway, whose damn book is it?

Indenting - The "right" thing to do is to line up chapter, section, and first paragraph headings on the left margin, and not indent the first line. Like this:

Right Way
This is the "right way to start a first paragraph, and a chapter heading.

My Way

This is how I have done it throughout the book I just think it looks less cluttered and easier to read.

Also, I'm supposed to put the titles closer to the text under them and I'm not supposed to use "so many" typefaces. (This is like telling an artist not to use "so many" colors in their painting.) My idea on the typefaces, was to use different ones for each of the eight Instincts, to underline their uniqueness. For instance, the typeface for Attraction Instinct has eyes in it - I just thought it was fun. As far as spacing goes, I think a little space between thoughts gives you a pause to digest them. Also, apparently, I numbered the pages "wrong". Silly me, I thought page one in a book was page one. Apparently it's not, it's page xii or some such silly thing. I didn't assign chapter numbers either, but do you really care what *number* the chapter is on Warrior Instinct? Add this to the fact that I have no credentials, and therefore no credibility as an author to the publishing industry, so why are you still reading this book???

But our Warrior Instinct dictates that right is right, and wrong is wrong. And those of us with Warrior Instinct have to keep doing things the "right" way, the way they've always been done.

Just for the sake of argument, let's look at a few things that were invented by people who *didn't* keep doing things "the way they've always been done" - electrical power, the computer, software to lay out books, and the printing press. I rest my case.

Semper Fidelis - Our modern military is the greatest achievement of our Warrior Instinct. Our military has its own structure of command, its own laws and police, its own clock, clothing, food, and code of behavior. Those of us with Warrior Instinct have been able to fashion a society, within existing societies, that serves as a place for our particular Instinct to thrive. But, even within the military there is competition. Air Force personnel may think they are better than Navy personnel, and the Marines may think they are better than everyone. Competition is standard operating procedure for those of us with Warrior Instinct.

Us modern Warriors are complex human beings, capable of taking someone's life without hesitation, and also ready and willing to jump into the line of fire, to *save* someone else's life without hesitation. The dividing line is who we believe is the enemy. Once our Warrior Instinct has been trained and given the green light, you had better have a continuous supply of enemies. If you run out, we can turn on the people who trained us. There is no on/off switch on this Instinct. Any one or anything can be a potential enemy.

The best soldier probably has a combination of Warrior and Worker Instincts. They would be extremely loyal, never question orders, and be willing to die for whatever they have been told is right and wrong. This combo would work well for enlisted ranks. The best officer combination is probably Warrior and Hunting Instincts. This person would have the mind for strategy and planning, and yet still a clear grasp of wrong and right.

Support Our Troops! - We say it as though it was an order. Those of us with Warrior Instinct don't question authority. We cannot love our country, be willing to die to protect our family, **and** think that our leaders are idiots. Those of us with Warrior Instinct have to accept the whole package, we can not allow ourselves to separate it out.

We Warriors live by slogans: "United we stand", "Love it or leave it", "Give me liberty, or give me death", "Second place is first loser", "Pain is weakness leaving the body", "Die trying", "No Fear".

Chief Executive Officer - John Pierpont Morgan created some of the largest corporations the United States has ever seen. He formed American Telephone and Telegraph (AT&T), General Electric,

Pullman, International Harvester, Western Union, Westinghouse, and United States Steel, which, at the time, was the largest corporation in the world. He controlled over 5000 miles of railroads, several banks, and insurance companies. His vast power and influence over the country's largest corporations, led him to be investigated by Congress. He forced Congress to create the Federal Reserve bank, by causing insolvency in several banks, and then bailing them out with his own money, just to prove to them that he was right.

Perhaps the most successful corporations are the ones run by those of us with Warrior Instinct, who see the competition as the enemy. Driven by our Instinct to wipe out the enemy, we captains of industry have run shotgun across the globe, through hostile takeovers, cornering the market, tying up distribution, and buying up stock.

Those of us with Warrior Instinct can be great to work for if we believe you are on our side. Or, we can be nightmares if we perceive our employees as enemies also. Consequently, wey may be so focused on our business, that we are not emotionally or even physically present at home. And if our partner complains, we may start to see them as an enemy, and the relationship may be toast.

Economic Growth - Much of the stability of our countries is based on money; specifically, the ability of our corporations to make a regular predictable profit from year to year. A country is considered healthy if the corporations in that country are seeing regular economic growth. But a continued rise in our profits is unsustainable. At some point, the market becomes saturated. This is the short-sightedness of our Warrior Instinct. Warrior

Instinct sees the small picture, the immediate issue at hand. And the immediate issue is more money.

At some point, all the stores are full of inexpensive consumable goods, and buying levels off. *Need does not grow evenly with profit.* It is a limited system, and will eventually plateau. But, those of us running corporations may continue to expect the same growth each year, and we will expand and borrow based on this assumption. And when our profits start to plateau, we blame consumers. We say consumer confidence is low. If profits are sluggish, we say the economy is sluggish, and then we start laying people off, closing stores and blaming consumers. Which creates a sluggish economy because people are out of work and have no money. Since our Warrior Instinct believes it's always right, it cannot be our fault. It cannot be that we over estimated our profit and expected too much. It always has to be someone else's fault. We are the architects of our own misery. Or, more specifically, our Warrior Instinct is the root of many of our own problems.

Monday Night Football - Competitive sports are one way for those of us with Warrior Instinct to be physically aggressive and not take lives. Sports offer us a socially accepted place to use all our skills. We can train, plan, practice and execute strategy, and we can have direct physical contact with other human beings, all in the name of good clean fun. Paintball courses and virtual reality combat games also fill this desire. They provide us with the opportunity to compete and vent aggressive behavior and win, in a secure controlled environment, without risking our lives.

Look at the similarities between those of us in professional sports and in the military: both have uniforms with insignia, both work as a team, both train and practice, we develop strategies to use against our opponents, and both are focused on winning. The only difference is, in sports, we aren't trying to kill each other. (With the possible exception of boxing and ice hockey.)

But those of us with Warrior Instinct can also satisfy our Instinct by simply watching sporting events, and even watching movies about people who fight and kill and drive fast. We can project ourselves into the situations we are watching, and, if we can identify with the side that's winning, it satisfies us. The same can be said for watching NASCAR, powerboat racing, motocross, tractor pulls, monster trucks, snow boarding, The Olympics, and even fishing. The underlying attraction is the same: our Warrior Instinct loves competition, doing it, watching it, talking about it, and fantasizing about it.

Virtual Firepower - Video and computer games are also ways to safely act out our aggressive behavior, without actually killing someone. The vast selection available, and amount of money spent annually on these kind of games, suggests that this Instinct is alive and well. Those of us with Warrior Instinct buy a lot of computer games. We buy bigger hard drives, more memory, faster processors, and better graphics cards. We are all about stronger, faster, and more power. We are the target market for new and innovative technology. We are cavemen with laptops.

An Error Has Occurred - The standard platform for computers which handle the world's business is a platform that is legendary for freezing, shutting down, and losing data. And yet, we will stand by a dysfunctional system and defend it to the death, rather than change loyalties. Hunting Instinct would make us switch at the first sign of unreliability. Hunting Instinct seeks the best, most efficient way to do something. Inventor Instinct would take it apart, and re-invent it to work better. But Warrior Instinct will stand by and defend a failing system to its dying breath. Those of us with Warrior Instinct prize loyalty over everything.

The flip side of this is, we are loyal to the death. It is the trademark of Warrior Instinct. Once we choose something, it's carved in stone. We are *always* right.

X-TREME - Those of us with Warrior Instinct live our lives by extremes. To take someone else`s life is an extreme thing to do. To be willing to lay down and die for something or someone is an extreme thing to do. To be willing to go to extremes is the true Warrior nature. And our language and behavior follows it. Right now we have Extreme Sports, Extreme Fitness, Extreme Challenges, and even Extreme Nutrition. This is the language of our Warrior Instinct.

The way we justify things is by extremes also. For example:

You, " I don't think I agree with our foreign policy towards Country X"

Warrior, " Well, maybe you should just go live in Country X if you love them so much!"

You, "I didn't say I loved them, or even agree with them. I just said I don't agree with **our** policy."

Warrior, "Well love it or leave it!"

In our Warrior Instinct world there are only two options: Our way (the "right" way), and any other way (the "wrong" way). There is no such thing as compromise or a third possibility. The way our Warrior Instinct justifies what it believes, is to compare it to some extreme opposite that is so obviously unthinkable, that you would have to be an idiot to believe in it.

The black and white belief system of our Warrior Instinct extends to all areas of our lives. If you don't think and act exactly like we do, then you may get labeled as being some kind of extreme opposite. If you don't do *one* thing the way we do, then in our mind you probably don't do *anything* the way we do. That makes you wrong, and it is our duty to point out to you how wrong you are.

Where our Warrior Instinct justifies with extremes, our Nurturing Instinct justifies with totalities - "Well *everyone* wants to be healthy, don't they?". Our Gathering Instinct justifies with fear. Our Hunting Instinct justifies with facts and statistics. Our Inventor Instinct justifies with logic. And our Worker Instinct justifies with feelings.

Words, words, words - Our Warrior Instinct burns through adjectives like gasoline. We have burned up words like, "Mega", "Power", "Total", "Accelerated", "Advanced", and "Ultimate". Right now "Extreme" is the most popular word we use to describe things. But, what is more intense and extreme than "extreme"? At some point, we will have to invent new adjectives or redefine already existing words in order to describe things in such a way that is satisfying to our Instinct.

Little Green Men - Those of us with Warrior Instinct live in fear. It is what makes us good Warriors. We live in fear of people and situations we can't control. The idea that there may be aliens communicating with us, or landing on the earth, is something that we would have no way to control. The best we can do is to deny, cover up, and destroy the evidence. Hunting Instinct would welcome the new information. Inventor Instinct would delight in the possibility of encountering a new species. But our Warrior Instinct is scared of anything it can't control. Even something as beautiful and non threatening as the crop circle designs that appear in the fields of southern England, are threatening to those of us with Warrior Instinct. We may even claim that *we* made them, in an attempt to control the situation. To have something that obvious, and unexplainable, is frightening to our Warrior Instinct.

The Devil made me do it - Some of our religious myths are based on extreme opposites, like a fight between good and evil, (or right and wrong). Some of us fundamentalist Christians even call ourselves, "Warriors for Christ". Indeed, evangelical Christianity would make no sense if there was no such thing as the Devil. What would there be to fight against? Sin and temptation wouldn't exist. There would be no point in us joining the religion. Fear is our prime motivator.

Our Warrior Instinct is also the origin of hate crimes. This can be seen in the relationships between religions. Extremist Muslims hate Extremist Christians, Protestants hate Catholics, and so on. And some of our religions are more Warrior-like than others. The Judeo-

Christian and Islamic religions, start more wars than the Buddhist, Hindu, and Taoist religions. Some of our long term conflicts that we are still waging on this planet are between religions that even follow the same text, just different versions of it. Rewriting a religious text in order to give credibility to our own personal beliefs is an act of our Warrior Instinct. Some of us will even try to rewrite history in favor of our own religions. There is nothing we will not try, in order to win.

INNOCENT UNTIL PROVEN GUILTY - Both sides of our laws are fueled by our Warrior Instinct. Police, Judges, and Attorneys are sworn to protect and serve the members of society by upholding the laws, prosecuting transgressors, and defending rights. Criminals use their Instincts to plot and execute crimes, and to outwit the police and courts. It can be a true battle of strategy and aggression. And, from looking at combined Instincts, you can see the origins of certain crimes. Our Warrior and Mating Instincts combine to commit crimes of passion. Our Warrior and Gathering Instincts combine to commit crimes of theft. Our Warrior and Inventor Instincts combine to commit crimes of intellectual property. Our Warrior and Hunting Instincts combine to strategically manipulate other people to commit our crimes for us. And those of us with pure Warrior Instinct simply blow someone's frigin head off.

A clever criminal with a clever attorney can commit murder and go free. When we do catch and prosecute someone with Warrior Instinct for a crime, we put them away with other angry Warriors, where they can learn ways to not get caught the next time. Or, they can learn new crimes to commit. The punishment for doing something wrong is short-sighted and ineffective, because it was invented by our Warrior Instinct, not by our Hunting or Inventor Instincts. Those of us with Warrior Instinct will typically leave ourselves a loophole to get out, since we don't like to bear consequences. We will build into anything we invent a way to escape the consequences. This makes for uneven punishments, and laws that can be manipulated to our own advantage.

Our Warrior Instinct is short sighted due to it's nature. The need for us to feel safe generates behavior to try to control other people. Yet the only ones who react negatively to this attempt to control, are other people with Warrior Instinct. Those of us with Warrior Instinct smoke pot. And those of us with Warrior Instinct put people who smoke pot in jail. Those of us with Warrior Instinct exceed the speed limit. And those of us with Warrior Instinct punish people who exceed the speed limit. Those of us with Worker Instinct don't smoke pot, and those of us with Hunting Instinct don't exceed the speed limit.

In other words, our Warrior Instinct creates more laws, and thus creates more criminals. It is the same as one group of Warriors building a bigger army, because another group of Warriors is building a bigger army, and one group stockpiling weapons inspires another group to stockpile weapons. Our Warrior Instinct creates conflict with its behavior towards other people with Warrior Instinct. It is our Instinct. There is always an enemy. There is always a need to prepare for conflict, because there is always conflict. Good luck trying to convince us

Warriors of this; we tend to see it the other way around. Of course we do, because it's an Instinct. Our Warrior Instinct creates conflict. It is our nature.

Look at the way we Warriors describe things we invent. " Law Enforcement." To en*force* the law. In other words, to force people to obey the law, by forcing them to pay the consequences if they break it. We set ourselves up for conflict from the start.

Department of Corrections - By its name, those of us with Warrior Instinct suppose that we can "correct" the way other people think and act. Because we know the *right* way to act, and these other people are acting *wrong*. But we cannot "correct" Warrior Instinct, which is why incarceration rarely changes people. When people with Worker, Gathering, Inventor, Hunting, and Breeding Instincts commit crimes and get imprisoned, they will change their behavior. But people with these Instincts are the smallest percentage of prisoners. They don't commit crimes. They have no reason to. They commit crimes by mistake, in moments of passion, or out of shear stupidity. Only Warrior Instinct tries to force things to happen and rebels against the control of other people with Warrior Instinct. Prisons are predominantly full of Warriors. And us Warriors have little or no success in forcing other Warriors to think and act differently. It is our Instinct. We would not be very good Warriors if we were that easily changed.

Warrior Instinct is an Instinct that can not be satisfied. There will never come a time when there won't be any more crime, because there will never come a time when there won't be one of us with Warrior Instinct trying to control someone else's thoughts and behaviors. There will never come a time when our Warrior Instinct will feel totally secure and safe. Like our Gathering or Nurturing Instincts, constant vigilance is the only way we know how to live.

No consequences - The main reason our Warrior Instinct is effective is because we don't look too closely at the consequences of our behavior. If a soldier stopped and thought about the family of the man they are about to kill, and what this loss would do to his wife and children, he may not pull the trigger. And if a fireman stopped to think about the personal danger he is putting himself in, rather than just rush into a burning building to save another human being, he probably would not go. What makes us people good at our jobs is also the thing that can make us hard to live with: we don't focus on consequences.

Not looking at consequences, unfortunately, can lead to not *caring* about consequences. If we apply this to our family life, relationships, or company policy, we can have big problems. Remember, there is no on/off switch for Warrior Instinct. Our Instinct can motivate us to do just about anything, and not feel any consequences. We cheat on our partners, lie to our families, steal, and kill. But those of us with Warrior Instinct may be confused when we get caught. We see the world only in terms of right or wrong, and we perceive ourselves as being on the side of right. Therefore, we could not possibly have done anything wrong. We may immediately seek to blame someone else for the whole thing. Sidestepping blame is classic Warrior behavior. If we cheat on our wife and feel guilty about it, so guilty we can't even look her in the eye, we will create a diversion. The most typical is to accuse *her* of sleeping around, and use that as a reason to end the relationship. Wrong or right, the point is we have to get her out of our lives so we don't have to look at our own consequences.

If we are in charge of armies or foreign policy, then the same behavior can repeat on a much larger scale with farther reaching consequences: "It's not *our* fault; we *had* to bomb their country. We *had* to kill them; we are right and they are wrong, so they had to die." But then we are greatly surprised and angry when those people strike back at us. Our Warrior Instinct makes us think that everyone thinks the way we do, or *should* think the way we do. And if they don't, then we will try to force them to think differently, or kill them. Either way, we still believe there will be no consequences for us, or *should* be no consequences for us, because after all, *we* were right!

NOTHING MORE THAN FEELINGS - Some of us with Warrior Instinct are often accused of being afraid of our feelings. Or we are accused of being insensitive and are sent to sensitivity classes to learn how to act differently. A true Warrior is not afraid of much at all, or we would be a lousy Warrior. Our Instinct has a trigger which shuts off feelings so that we can do our job. All the sensitivity training in the world may not be able to change an instinctual response. Our Warrior Instinct can also make some of us uncomfortable with intimacy. We cannot feel compassion for someone we are about to kill. This is instinctual behavior. We are born with Warrior Instinct. We don't learn it. It has not been conditioned into us by the society we live in. You can see it in every country on earth, in every culture, regardless of economics, climate, or history. Those of us with Warrior Instinct do the same identical behaviors, and have the same identical attitudes. It is an Instinct, pure and simple. Some of us act like we were born angry. Perhaps, what we were born with, was actually Warrior Instinct.

Some of us are also accused of being afraid to commit to a relationship. Again, if we have Warrior Instinct, we are not afraid to commit to anything. Warriors are probably the most loyal and committed people on the planet. How much commitment did it take for Japanese pilots to climb into their planes and willingly fly into their own deaths at Pearl Harbor? Commitment is not the issue; *what* we are committing to is. There are further problems because all eight Instincts are committed to different things. Of course they are. That's what makes those Instincts effective. The problems start when we point our fingers at each other and complain that the other Instincts aren't committed to the same things that *we* are, and that's not acceptable to us Warriors.

And, remember, we now live much longer than we ever have. Relationships between people are much different than they were for the first 3.5 million years. But the Instincts haven't changed. Those of us with Warrior Instinct are still not instinctively motivated to hang around the camp. Our duty calls us to be mobile and ready at any time. The concept of having a relationship forever, and living to be an old person with the same spouse, is completely foreign to our Warrior Instinct. It is the desire of our Nurturing Instinct.

Some of us Warriors can be uncomfortable with intimacy. After we have sex, we may move to the other side of the bed or get up immediately and take a shower. We are not into touching and cuddling. Nurturing Instinct desires touching and cuddling. Some people with Nurturing Instinct would almost *rather* have the touching and cuddling. Unfortunately, since Warrior and Nurturing Instincts appear to make us attracted to each other, intimacy can suffer, due to this difference in our Instincts.

Crips and Bloods - Street gangs are perhaps the purest form of our Warrior Instinct still left over from Primitive times. Being a member of a gang gives us an identity and a feeling of self-worth. It is a modern-day primitive tribal bond. We will do just about anything for another gang member, and we usually have some kind of formal organization to our groups. Roaming through the concrete jungle defending what we believe is our territory, is our primary occupation. Frequent skirmishes with other gangs provides us with the opportunity to practice our skills at combat. We wear our gang insignia with the same pride and arrogance as a soldier wears a uniform, and, just like a soldier, we value loyalty over all else.

Trying To Represent - Hip-hop music is the music of the streets. It is born out of the day-to-day lives of those of us living in poverty in the United States. Since Warriors don't die in battle in the great numbers they used to, there is a large number of us with Warrior Instinct left hanging around the camp with nothing to do. Our music becomes the way we posture and brag about our strengths and superiority. If you listen to the music of the streets, you will hear our loud angry rhymes about how we are the best, and our rhymes are the best, and how other people don't compare. It is the posturing and boastfulness of us 21st Century Warriors who cannot get respect by holding a weapon. Even though, some of us songwriters do cross over into being gangsters, and even die from being shot by other 21st Century Warriors. But then we just make another CD about being shot, or our friend who was killed. And we will write new songs about how great our friends' rhymes were, and how everyone else sucks by comparison.

If I hit you, and you don't get back up, that means I'm right - I say you are wrong, You say I am wrong. Then I say you are full of shit, and you say I'm full of shit. Then you hit me. Then I hit you. And we keep on hitting each other until one of us is still standing. The one who is still standing is right, and the other one is wrong. Right? This is pure Warrior Instinct. Might makes right. If I physically beat the crap out of you, then that means that I'm right. You can see this mentality being exercised in bars all over the world. And it is the basis of our military strategy. The one with the most people left standing wins, and that means that they were right and the others were wrong.

You can also see this in the courtroom - those of us with Warrior Instinct sue other people to prove we are right. We believe that if we win the case, and the court forces you to pay us money, then that means we were right. In some countries, if you have enough money, you can hire a team of attorneys that can spin almost any case in your favor. And, for a Warrior, it's all about the principle, not the issue. Us Warriors needs to be right, even if we are wrong. We get our needs met by force, or threat of force. This can also include forcing someone to give us money. We have to be right. *What* we are right or wrong about is not as important. The important thing is, that we were proved to be right in the end.

It Must Have Broke Itself - True story. The mechanic made a mistake installing the front strut on the car. When he took it out for a test ride, it stripped out the threads and broke the strut housing. Rather than admit that he was wrong, he blamed the car. His story was that after he replaced the strut, the strut housing somehow became defective all of a sudden, and broke itself. (The car is to blame). Then he went into an assessment that this particular car (a Volvo 240) was not built very well to begin with, and I should really be driving a different kind of car. (I am to blame). Then he proceeded to explain how it was actually better that this was broken, since now I had a good reason to go buy a different car. His thought was that he actually did me a favor by breaking the strut housing. Us Warriors have to be right, even in the face of clear physical evidence that we are wrong. This is pure instinctual behavior.

World's Best! - In the early days of the United States, our immigrant ancestors risked everything to travel across the ocean with their hopes and dreams. They left behind family and history to build a better life in the new world. Our Warrior Instinct can make us believe that no matter what it is like in the new world, it *has* to be better than what we left back in Europe. It cannot have been a mistake to leave our homeland. It *had* to be the right thing to do. Life in Europe has to be worse, in our minds, than what we have built here. This can lead us to believe that *everything* we have in the United States is better than anything anywhere else.

If one thing is better, then *everything* is better. If we escaped from living under a tyrannical government back in Europe, and now we live free, we may boast that we are *more free* than anyone else on the earth, and that people in other countries *wish* they lived in the United States, and had the freedoms that we have. We may not even know what life is like today back in Europe, and we don't care; we're Americans now. We may not even be able to list what freedoms we have that people in European countries don't have Doesn't matter; our Warrior Instinct is absolute. In our minds, the United States is the best place to live in the world!

This can often spur us to claim that the products we manufacture in the United States are the "World's Best!" This is not a claim based on a comparison of every similar product manufactured in every country. This is a boast born out of our Warrior Instinct that *anything* and *everything* we make in America is always better, simply because it was made in *America*.

Road Warriors - I am the one riding your bumper in the morning rush hour, and you are the enemy. It's not that *I* got up late, or I didn't leave home soon enough to avoid the traffic jam. No, it's actually *your* fault that I am late because you are in my way. And if I have to speed and get stopped by the police, then it will be *their* fault that I am late. If I have to wait for you to get onto the interstate, as soon as I get up to speed, I will immediately pull out into the fast lane and go around you, and speed up to where I would have been, if you had not delayed me. In my mind I have to make up for that 1/100th of a second immediately, and I will probably not use my turn indicators, either. *I* know where I'm going, and screw you anyway. My Warrior Instinct makes me park in the handicapped spaces in front of convenience stores, because I'm "Only running in for a few things, and to hell with that restricted parking crap!" My Warrior Instinct believes laws and rules are for other people. My Warrior Instinct doesn't care what the frigin speed limit is. I want to drive race cars on the highway. I believe that low performance cars are for low performance people. I buy radar detectors, and drive as fast as I want to. In my mind, laws are for other people.

The problem is, my Warrior Instinct creates the laws and rules I use to try to control other people's behavior. But my Warrior Instinct doesn't like *other* people with Warrior Instinct trying to control *my* behavior.

Bigger is better - Those of us with Warrior Instinct will buy the biggest truck we can find with all the options. Then we will spend thousands of dollars customizing it. We will remove the brand new tires and wheels, and install custom wheels and oversize tires. We will put on running boards, cab lights, winches, and roll bars, even though the heaviest thing we will haul in it will be an 40 pound golf bag. And, we will wash and wax the thing at the very least sign of dirt.

A large house, an expensive car, membership in certain organizations and clubs, and a strong stock portfolio, are obvious signs of a successful person. These things provide us with our feelings of self-worth. If we have Warrior Instinct, we will buy the biggest house the bank will lend us the money for. We will lie on our application and over-inflate our income to try to squeak out as much credit as we can. Our Warrior Instinct is the origin of lying. We lie for two reasons: to make more out of something than there really is (the flashy show), and to avoid the consequences of our behavior.

The Women's Movement - Women with Warrior Instinct burned their bras in the 60's as a clear message that they would no longer set themselves up as sexual objects for the pleasure of men. Push-up bras are not designed by men who force women to wear them against their wills. Push-up bras are designed by women with Attraction Instinct to make themselves feel more attractive. Our Warrior Instinct doesn't motivate us to be attractive to the opposite sex. Our Attraction Instinct motivates us to be attractive to the opposite sex, and our Nurturing Instinct motivates women to be mothers and wives.

The Women's Liberation Movement was a movement by women with Warrior Instinct to change the way women with Attraction, Mating, and Nurturing Instincts think and act. Essentially, it sought to make them think and act more out of their Warrior Instinct, and to try to get males with Warrior Instinct, to think and act differently. Women's Liberation spawned the Women's Movement, and fostered Women's Studies in colleges and universities. Now, the focus is on educating young women *away* from their Breeding Instincts and *towards* their Warrior Instincts. And, to educate them and young males as to how males have subjugated and controlled women in the past, and how to not let that happen any more.

But again, females with Hunting, Worker, Gathering, Breeding, and Inventor Instincts do not feel oppressed and want to fight back. Only Warrior Instinct desires to be in control. And men with Worker, Inventor, Hunting, Breeding, and Gathering Instincts don't seek to control women. Only men with Warrior Instinct desire to be in control. Again, Warriors only fight with other Warriors. In this case, it is female Warriors against male Warriors. But, the Warrior Instinct is so absolute, that it believes, "If it's *my* issue, then it's *everyone's* issue!"

Cave People With Cell Phones - Political Correctness attempts to make us use different words, in order to change the way we think and act. This is motivated out of our Warrior Instinct, and clear ideas about what is wrong and right. It is directed towards other Warriors. Warriors only fight with other Warriors. Those of us with Worker Instinct don't care about words; however you want to be addressed is fine with us. We regard it all with mild amusement, change the words, and go on. As usual, the people that we Warriors are trying to change, are other Warriors. And they won't change, because they believe they are right. Neither of us will budge. We both think that we are right. Warriors trying to change other Warriors is the basis for most of the conflict we have as a species.

Civilized conflict resolution - For the first 3.5 million years we fought hand to hand. In the last 10,000 years, since we became civilized, we fight very differently. From Roman Legions to the Revolutionary War, well-dressed rows of soldiers marching in unison and dropping to their knees to fire at other rows of well-dressed soldiers was considered the "Gentleman`s" way to resolve conflict. As recently as the American Civil War, two opposing groups of Warriors would meet on an open field and face each other and shoot until one side clearly had more people left than the other. And then they would be declared the winner.

Since the Revolutionary War, the technology of fighting has changed dramatically. Europeans started hiding behind trees and rocks and shooting at each other. And now we can sit on a ship miles out to sea and launch missiles at each other without ever getting in harm's way. Two great armies don't have to meet face to face. Our Warrior Instinct has developed better and better weapons that allow us to inflict damage on our enemies from greater distances without dying ourselves.

Consequently, we lose fewer and fewer soldiers during conflicts. According to the Center For Defense Information and the Department Of Defense, approximately 214,000 soldiers died during the American Civil War. In World War One, the number of soldiers killed was close to 53,000. In World War Two, the number of dead was over 290,000. But during the war in Vietnam, the number of dead fell to about 47,000, due to advanced technology. And in the Persian Gulf War in 1991, less than 300 US soldiers died. This advancement in technology has also dramatically shortened the length of time we are involved in fighting. Consequently there are now many more of us with Warrior Instinct left hanging around the camp.

CONFLICT ISLAND - For all of our advances in technology and education, we still try to resolve conflict in the same way. If one group of us Warriors threatens another group, we attack each other, destroy dwellings, and try to kill as many of each other as we can. This approach has not changed in almost four million years. The idea is, that if we do enough damage to the other group, and kill a bunch of them, they will leave us alone and not do the same to us. And, for as long as we have been on the earth, we still believe that this is the way to provide safety. And when we do attack, it disrupts the food, water, shelter, and safety of the people in the other country. Those people get caught in the conflict and die, services are interrupted, and buildings must be rebuilt. It is inefficient and expensive for all concerned.

The *people* of one country have never declared war against the *people* of another country. It is usually a handful of us with Warrior Instinct, who can't get along with other Warriors, who declare war. We declare war *in the name of the people* of our country. If we feel like someone is our enemy, then we may believe that they are *everybody's* enemy. Warriors only fight other Warriors. Those of us with Gathering Instinct don't fight; it serves us no useful purpose. Those of us with Worker Instinct would rather give in than fight. Those of us with Hunting Instinct will look for a more cost-effective resolution. Those of us with Nurturing Instinct will avoid any kind of conflict at all. And the ones with Inventor Instinct will look at the bigger picture and seek long-term solutions. Those of us with Warrior Instinct only go to war with other people who have Warrior Instinct. And yet, it is the ones with Worker, Gathering, Breeding, Hunting, and Inventor Instincts who suffer the greatest losses of life and property. And yet, those of us with Warrior Instinct usually find ways to make the others fund our wars, whether it's by forceful taxation or scare tactics about their safety.

But, no matter how many or how few soldiers die in battle, there are always civilian casualties and destruction of homes, schools, roads, and public utilities. If only there was a way those of us with Warrior Instinct could fight each other without killing innocent people, and without destroying the infrastructure of countries.

Perhaps if there was a remote uninhabited island that was set aside to only serve as a place for our armies to meet and resolve our conflicts, then we wouldn't have to spend so much on security and defense, and only us with Warrior Instinct would die, not innocent civilians. We wouldn't have to keep rebuilding what we keep destroying. Just a thought.

COMPATIBILITY

SCREWING AROUND - The enemy is someone or something outside of us; it cannot be us or anything that we are doing. If we cheat on our wife and she leaves us, *she* will be the one who is at fault: she wasn't worth being faithful to anyway, or she didn't keep us interested, or she was probably cheating on *us* too, and so forth. And, if our ex-wife then starts seeing some new guy, then the new guy becomes the reason why she left, not the fact that we cheated on her. Even if she meets this new guy *after* she leaves us, the whole thing can still be the new guy's fault. Those of us with Warrior Instinct cannot accept consequences or it may interfere with our abilities as a Warrior.

I'm outta here! - Some of us leave relationships at the first sign of trouble, not giving them time to work out. A primitive Warrior or Hunter would intuitively know that a stationary target is a dead target. The way to stay alive is to keep moving. Remember, this is not a conscious rational thought, but rather an knee-jerk instinctual reaction that is triggered by a situation. Reacting to this Instinct has kept us Warriors and Hunters alive for millions of years.

A few good men - Trying to have an intimate relationship with one of us who has Warrior Instinct may be impossible. Some females are attracted to males who are the big strong Warrior types. This could be an instinctual leftover from Primitive times, when these males were the ideal one to impregnate them. But try raising children, balancing a checkbook, and planning for the future with someone who is constantly on guard for enemies. In the United States Marine Corps they have a saying which underlines the Warrior Instinct: "If the Marine Corps wanted you to have a wife, they would have issued you one."

Of course I'm right - Ever been in a relationship with someone who just *has* to be right all the time? And even if you prove to us that we are actually wrong, we won't admit it! We will do anything to get out of the spotlight, including bringing up something that you did wrong in the past, as though that cancels out what we did. For some of us with Warrior Instinct, it appears that it is more important to be right than it is to be happy. We Warriors have to be right. In battle, if we aren't right, we might end up dead. And so we believe that giving advice is being helpful. We don't understand the concept of supporting someone in finding their own answers. We tell you straight out, "Just do what I say, and that will fix it." To someone with Hunting or Inventor Instinct this may make us look like a know-it-all, or a jerk who doesn't think other people have a brain too.

Some of us with Warrior Instinct seem to need a certain level of conflict in our lives. It doesn't have to be physical; it can also be emotional or intellectual. This is not a conscious choice to cause problems. We are simply driven by our Instinct to win fights. So, if there isn't a fight happening, we may instinctually start one. It is what we do. We can often look angry. It is because we see everything in the world as right or wrong. And, all the stuff we see as wrong gets on our nerves. We believe it is our duty to point out the wrong stuff and show you how to do it right. This is how we cause conflicts, especially if we try to tell another Warrior that they are doing something wrong.

Some of us will even argue over something we know nothing about just to win the argument. In our mind, if we win the argument then that means we are right. It is of little importance what the issue is; the point is to come out the victor. We Warriors absolutely have to be right. We will even rewrite history if we have to. Our need to win an argument knows no boundaries. If you stumble over a word, or mispronounce a word, we will seize upon that as a clear indicator that *everything* you are saying is wrong and can be dismissed.

Face it, those of us with Warrior Instinct may not be cut out to have long term relationships. For most of our history on this planet, we only lived to be 25 years old. Look at 25-year-olds today. Up to that age we can pretty well get along with anybody. By the time we are 30, we start to get stuck in our ways. We form clear ideas of wrong and right, and good and bad. In Primitive times, odds were that a Warrior would get killed in battle, or stomped by a woolly mammoth. And a true modern Warrior lives the same way: We are here today and may be gone tomorrow. We live for the thrill of the battle, even if it is only over the remote control for the TV.

Partners

Warrior and Hunting - See Hunting and Warrior - page 31.

Warrior and Gathering - See Gathering and Warrior - page 53.

Warrior and Warrior - One of us would have to take the upper hand in some things and allow the other one to take the upper hand in others, or else we just might kill each other. The territorial nature of our Warrior Instinct is going to cause space issues in the relationship, specifically: my space and your space. We will have tension over money, possessions, time spent together versus time spent apart, and just about every other small stupid argument we could possibly run into. But, the anger can be passionate and we could have a very intense sex life based on venting our aggressions in a healthy manner towards each other. Or, we could be abusive and physically assault each other. The relationship can disintegrate into a standoff over who is right and who is wrong.

We are animals. If a bear discovers a stream full of fat salmon to eat and another bear comes along, do they shake paws and share the fish? No, they fight over it. We are more like the other animals than we are different. Those of us with Warrior Instinct fight with others who have Warrior Instinct. It is our nature. It is our Instinct. Warriors fight.

Warrior and Worker - Our Warrior Instinct promises security and safety, because feeling unsafe and insecure is our normal state of being. It is the thing that makes us good at our job. Someone with Worker Instinct can see this as a guarantee that they will not have to worry about things changing, and generally not worry about anything threatening them, ever. Our Warrior Instinct, however, protects by controlling. The only way for us to feel truly safe, is to eliminate anyone who could potentially threaten us. Obviously, we aren't going to kill the ones we love. But, we will demand certain restrictions, and adherence to a code of behavior, so we can feel in control. In essence: "I love you and want to protect you, so please crawl inside this box, and stay there so I know that you are safe. It's for your own good. And it makes me feel safer, because I will know that you are contained." Those of us with Worker Instinct will willingly crawl inside the box. In our mind, other people are in control of our lives all the time anyway. And at least we know we will not have to worry about things changing.

To those of us with Worker Instinct, a familiar restricted existence is usually more attractive than an unknown freedom. We may grumble a little, but we will let the one with Warrior Instinct tell us what to do, pretty much every time.

This relationship does not change, from an individual level, all the way up to a global level. Those of us with Warrior Instinct get our power from people with Worker Instinct. Workers, who are scared of the unknown, will elect us to take charge of things. On a personal level, this can be a controlling, and potentially abusive relationship. On a much bigger level, this is how democracies turn into dictatorships. Those of us with Worker Instinct don't want the responsibility of running things. We will gladly turn over control, so we don't have to worry about the day-to-day details of management. Remember, we don't like to own the business; we just want to put in our time, and collect our paycheck. We will happily let people with Warrior Instinct do whatever they want. Those of us with Worker Instinct will tolerate enormous amounts of abuse and neglect in exchange for not having to be responsible for the livelihoods of others.

Our Warrior Instinct is also progressive. It does not waver or retreat. Warrior Instinct will take more and more, and impose greater and greater restrictions in order to feel safe. In the end, those of us with Warrior Instinct will start to dictate to those with Worker instinct, and the Workers will have no power or leverage to disagree. In short, democracy rule, or rule by the majority of the people, doesn't work for very long because the majority of us have Worker Instinct. We don't want the responsibility, and there are always a few Warriors hanging around who will be more than willing to take control of the situation. Those of us with Worker Instinct will vote for Warriors every time.

Warrior and Inventor - See Inventor and Warrior - page 140.

Warrior and Attraction - See Attraction and Warrior - page 162.

Warrior and Mating - Potentially explosive and deadly. Those of us with Warrior Instinct like partners who take action. But we don't like partners who screw around. Exciting for the one with Mating Instinct, but frightening when the initial infatuation wears off.

Warrior and Nurturing - The natural biological partners. However, trying to force us to stay together forever can cause problems. Remember, for most of our history we only lived 20-25 years. Any attempt to live together longer may be a doomed relationship that will be difficult at best, and deadly in it's worst. These are the relationships where spouse abuse occurs.

These relationships are full of miscommunication, uneven standards, and one-upmanship. People with Warrior Instinct and people with Nurturing Instinct in a relationship, basically have three choices. Split up, fight your whole lives and slowly kill the intimacy between you, or one of you gives in and you choose to live with an uneven relationship.

If we have Warrior Instinct, we can live in a ditch; we often have to during war time. We will not instinctually grab a coat if it is chilly outside. Surviving hardship is simply a test of how good of a Warrior we are. This is instinctual. But our Nurturing Instinct will see this as some defect in our thinking, because people with Nurturing Instinct would instinctually grab a coat. And so the judgment starts, "You are so dumb; you have to be reminded to wear a coat when it's cold. You're just like a little kid." And the Warrior responds, "You are such a pain in the ass. I'm no kid, I know when I need a coat and when I don't. Get off my back."

Our Warrior Instinct argues what it believes is the right and wrong thing to do, and our Nurturing Instinct argues what it believes is the good and bad thing to do. We both think we are talking the same language, but we aren't. Good and bad are judgments, and right and wrong are beliefs. Those of us with Nurturing Instinct judge the beliefs of people with Warrior Instinct and tell them they are bad. Those of us with Warrior Instinct see the behavior of people with Nurturing Instinct as wrong. The actual issue we are fighting over tends to get lost, and it becomes a battle of wits. Both of us feel like we are right; both of us think the other is just being stubborn and stupid. This behavior is instinctual. It will not go away. Our Nurturing Instinct believes it can educate Warriors away from their violent natures. We have been civilized for about 10,000 years. Has it worked? Are we less violent today than we were even 100 years ago? Do we have fewer conflicts raging? You might be thinking at this point that these sections on partners are overly negative. Ask yourself, how many happy functional relationships do you know of personally? Perhaps these are the reasons why.

(Also see Nurturing and Warrior - page 208.)

WARRIOR INSTINCT

IN A NUTSHELL:

Warrior Instinct compels us
to focus on security.
It drives us to control other people`s
thoughts and behaviors.
It makes us territorial and think in terms of
possession, ownership, and protecting
the things we believe are ours.
It makes us see things as wrong or right,
black or white, friend or foe.
It desires to compete and win.

WARRIOR INSTINCT IS SATISFIED WHEN:

We feel we have control over
people or situations.
We have been proven to be right.
We win.

WHEN IT ISN'T SATISFIED:

It uses force or threat of force.

GIVEN UNLIMITED RESOURCES:

It could destroy the species.

WORKER INSTINCT

For millions of years, we didn't have jobs where we earned money and stores to spend the money in. We didn't have to pay for housing or water or food or security. We didn't work. We didn't have to. Food grew all around us, streams and lakes glistened with pure water, trees and caves gave us shelter, and none of it cost us a damn thing. In the last 10,000 years we have changed the way we live dramatically. We invented money, and we reinvented ourselves as a new type of human being whose life depends on making money. If you tell us that we have to show up on time and put in 40 hours a week and run a machine which stamps out plastic forks so we can feed, clothe, and house ourselves and our families, the Worker Instinct is the Instinct that will make us do it. For the first 3 million years or so, no one punched a clock or took home a paycheck. The factory worker, the assembly line worker, and the common laborer are all the inventions of our modern Warrior Instinct.

Our Worker Instinct not only motivates us to bring home the bacon for our families, but it also makes other people wealthy. Until we invented civilization, there was no such thing as the middle man. If we were hungry, we found food. Nobody gathered our food for us and charged us for the service. We did it ourselves. Nobody built a cave for us, nobody fetched water for us, and nobody stood guard over our family. We did it all ourselves. Today these things are done for us by others, but we have to pay them for it. By inventing the middle man, we created something that we call "Working for a living".

Worker Instinct may be the most common of all the eight Instincts. "Blue Collar" workers make up the largest demographic group on the planet. Those of us with Worker Instinct get up every day, go to work, and do the same job, whether it's standing at an assembly line or sitting in a carpeted cubicle under fluorescent lights. Societies simply could not operate without us. Our labor creates the goods we trade, and that trade generates the wealth that makes nations stable. But, we Workers can be either the happiest or the saddest of people. The deciding factor is if we feel appreciated. We just want to feel like we are valuable. Those of us with Worker Instinct also desire security and predictability. If you can provide this to us, we will work faithfully for you until we drop.

As long as our paychecks don't bounce and we can afford to buy a few luxuries now and then, we will continue to show up every day and do our job. People with Worker Instinct are reliable, dedicated and committed. We take great personal pride in what we do. We Workers step up to the plate and pull our own share. We rarely complain to the boss, but will commiserate with each other for hours over a cold beer at the end of the day.

With the advent of the Industrial Revolution, people like Henry Ford created ways to mass produce goods to flood the market and provide things to ordinary people on a large scale. The other thing that he created was the meaningless job on the assembly line. Sure, we can take pride in the car we are building. But if we can't afford to buy it, and we can be replaced at any time by some other person who can be trained to do the same job, then where is our personal feeling of self worth? The answer, at least for us with Worker Instinct, lies in our families.

Those of us with Worker Instinct raise emotionally healthy children because we don't obsess over them. Worker Instinct is actually better at Nurturing than Nurturing Instinct, because Worker Instinct doesn't see it as a job. It's just part of the flow of our day-to-day lives. Those of us with Worker Instinct are committed spouses and reliable concerned members of the community. We take pride in neighborhoods. We feel proud to be a citizen of a nation, and we will rise to defend it whenever we are called to do so. Worker Instinct breeds patriotism; a patriotism based on a shared heritage of scraping away at the dirt to build our lives up from scratch. We don't feel the angry, self-righteous patriotic pride born from our Warrior Instinct. We are sentimental. We are proud of what we have built and what we have. We aren't inspired to boast or compare or shove it down someone else's throat. We remember the past and look to the future. But we don't threaten or force people to respect us. We just get up in the morning, look around at the good things in our lives, and put our shoulder to the wheel for one more day.

Our Worker Instinct tends to see the bigger picture, like Inventor and Hunting Instincts. It believes that we have little or no control over our destiny. Those of us with Worker Instinct don't sweat the small stuff, or even the big stuff. We take everything with a grain of salt and let things go their own way.

The civilized world would grind to a halt without our Worker Instinct. Our Warrior Instinct will not motivate us to pick up the garbage. Our Nurturing Instinct will not move us to clean the rest rooms at the mall. Gathering Instinct will not spur us to clear tables and wash dishes at a restaurant. Every society runs because there is a large group of us who will do whatever it takes to feed our family. It can be seen in every country around the world. The best-run companies are probably companies that have someone with Inventor Instinct for a boss and people with Worker Instinct doing the work. Inventor Instinct is project oriented; it sees the overall big picture. Worker Instinct is task oriented; we don't want to own the company because we don't want the stress or the responsibility.

Those of us with Worker Instinct generally will not strive to be famous. We don't typically seek publicity or attention. We may even talk ourselves out of our own success. We take very few risks and only ones that are carefully calculated. Our idea of success would be to win the lottery.

Those of us with Worker Instinct believe in the concept of talent. We believe that someone or something made other people more talented than us. We may even say it is a gift from some creative force. But if you point out a talent that we have, we will down play it. We are uncomfortable with the attention. We would rather just show up, put in our time, and then go home.

The very structure of our day-to-day lives is built on the desire of Worker Instinct for predictability: the 40 hour work week, five days on and two days off, guaranteed vacations, predictable pay and benefits, and the possibility of regular raises. This is one place where someone with Warrior Instinct knows how to talk to people with Worker Instinct. If they want to get elected, they don't campaign on a platform of big sweeping changes, because change scares us. They campaign on a platform of securing jobs, and a promise of working less and getting paid more.

PRIMITIVE

Every species has its worker bees. The basic Instinct to pitch in and help out contributes to the well being of the whole tribe. This means doing the manual labor of cutting wood, plowing, digging, hauling water, and so forth. The rewards are a predictable, well-organized living environment, and a sense of belonging to a group and being appreciated.

The Worker is the backbone of the community. Workers literally built the ancient world. An Inventor probably designed Stonehenge, but Workers showed up each day to drag the stones into place. Workers are the largest demographic of any hive. They are the strong arm and the bent back that gave us our history and our greatest marvels. Their labor built the pyramids of Egypt, the Great Wall of China, the Taj Mahal, and Notre Dame cathedral.

MODERN

Modern Workers are still the backbone of civilization. We are the ones who work the hardest, take home the smallest share, worry the least, carry the biggest burden, and yet somehow, always have time to bounce our children on our knees. Those of us with Worker Instinct may well be the most content people on the planet. The things that make us content are predictability, our long term dreams, regular vacations, and family. Where people with Nurturing Instinct talk about the importance of family, those of us with Worker Instinct instinctually live it. We can make reliable spouses and have long-term rewarding relationships. We raise healthy children because we don't pressure them with unrealistic expectations. We take few risks, maintain a sense of humor and sleep well at night.

We know we are just cogs in the wheels of progress. We readily accept our place in society, and put our shoulder to the wheel. But since we see the bigger picture, we don't stress ourselves out with concerns that would drive a Warrior, Nurturer, or Inventor crazy.

TYPICAL BEHAVIORS

THE BEST FRIEND - Those of us with Worker Instinct are usually lifelong friends to those who get to know us. We are the ones who will actually give you the shirt off our back. We are motivated purely out of the feeling that we like you. We may not know what to say or do in a given situation, but we will stay with you until you get it figured out. We offer support, not advice. We make no demands and are happy just to be your pal and hang out.

THE RIGHT HAND MAN - Worker Instinct creates the best assistants you can get. We make excellent secretaries, operating room nurses and dental assistants. We feel self-worth knowing that the boss couldn't get along without us. We will pass up promotion and work an entry level position until we retire. But we will be an indispensable part of any operation.

To have us close to you in your daily life means not having to worry about a lot of details. We are usually the ones who can tell you phone numbers off the top of our heads. We remember how many bags of mulch it took last year to do the flower beds in front of your house. We take the same road home every day, and make sure we stop for gas and milk. We are organized without being obsessive about it like Gathering Instinct. We just go about our business, and everything gets done.

THE BEST DAMN PANCAKES YOU EVER HAD - We are the ones who take pride in the small things in our lives. We are the ones who have hobbies and collections. We derive joy from building a collection up over time. We like to have shelves in our houses with small objects on them, and each object will have a story attached to it. We take pride in our flower beds, sleep under our grandmother's quilt, and have Tupperware boxes crammed with recipes. We are the ones about whom people will say, "You should try her pancakes sometime!" We like things that are tried and true. We like recipes that have been handed down from generation to generation. We value tradition and predictability.

I'LL HAVE THE USUAL - When we go to a foreign country we will eat at a fast food place we recognize from our own country rather than try anything locally owned. It is because we know what we are getting. It's the predictability that attracts us. Our bartender knows what we want and can set it up as we come through the door, and we will be happy. We get the same haircut we have had since high school. We listen to classic rock stations and probably still have the first album we ever bought. Our motto is: "If It's working for you, don't fix it." Typically, those of us with Worker Instinct will usually choose a familiar way of doing something over a new and unknown way to do something.

RANK AND FILE - Unions are full of Workers. This sounds like an obvious statement, but why would we be motivated to join a union? Job security. We want to know there will be work tomorrow, and next month, and next year. We don't want to be overworked, and we want to have our benefits. We want to make sure we get our fair share and are not being taken advantage of. We want a regular cigarette break and a guaranteed lunch. And if we have to work extra, we want to be compensated for it. But most of all, we don't want to have to worry about our job. We will gladly pay dues to someone else to do our worrying for us. We don't want to own the business. We don't want the responsibility. We just want to put in our time and get our check.

Our Warrior Instinct, however, often motivates us to become labor union management. Our Warrior Instinct makes us pressure people with Worker Instinct to join our unions. We are demanding and confrontational, and try to force employers to meet our needs. We see corporate management as the enemy, but we will not start our own corporations and run them the way we think they should be run. The security would satisfy our Worker Instinct, but our Warrior Instinct would lose its enemy. We cannot be our own enemy, doesn't work.

FORD OR CHEVY - Those of us with Worker Instinct have fierce loyalty. We will only buy certain brands and support certain sports teams. We may only drive the same brand of truck our grandfather drove. We may only drink Budweiser, and never Miller. If we live in Minnesota, we would not be caught dead in a Green Bay Packers coat. And, if we live in a flood zone and the river washes our house away, we will rebuild; even if the federal government will pay us to build somewhere else, we won't do it. "If it was good enough for my Dad, it's good enough for me!"

DON'T FORGET US LITTLE PEOPLE - Worker Instinct creates feelings of insignificance. If a friend of ours becomes successful, we may believe that we will be forgotten about. This is a deep-seated fear based on our belief system, not on reality. This is sometimes followed closely with the statement, "I knew them when ..." In this way, at least we can feel somewhat important - even if it is only briefly - because we knew this person before they became successful.

THE "MEANT-TO-BE" THING - Some of us with Worker Instinct desire predictability so much that we like to believe there is an all-powerful something or someone watching out for us. We may call it "God", or if we aren't religious, "The Universe". Some of us Workers may go so far as to believe that many of the details of our lives are controlled in this way. For example: We see an ad in the newspaper for a washing machine and call the number and it's busy. They try to call it a few more times for the rest of the day, and a couple of times the next day, and it's still busy. We may say, "I guess maybe I'm just not supposed to get this washing machine; it's just not a "Meant-To-Be" thing. The assumption is that "God" or "The Universe" is watching out for us, knows that this is not a very good washing machine, and is protecting us by making the phone be busy every time we call. The reality of the situation could be that this is actually a very good washing machine and an absolute bargain, but there are two teenagers who live at the house which is selling it, and they tie up the phone line constantly talking to their friends. Some of us will not usually push a situation to find out the truth; we prefer to believe that an unseen protecting force is directing our lives. It's a philosophy of living that is very comforting to some of us. Someone with Warrior Instinct, on the other hand, would call the house every hour until they see the damn washer for themselves. We Workers give in quickly.

Worker Instinct makes us truly believe that we have no control over parts of their lives that we actually do have control over. In comparison, Warrior Instinct makes us believe that we have control over things that we actually don't have control over.

SHIT HAPPENS - Having said the above, some of us Workers can believe in this outside control of our lives so much that we often see ourselves as victims. We may frequently use the word "They" when we talk : "They won't let me ...", "They expect me to ...". In our mind, this mysterious "They" are responsible for most of the problems we have to deal with. Some of us also like to use the word "Society": "If only society would ...", "What's wrong with society is ..." and so on. We may perceive society as everyone in the world, except ourselves. When negative things happen we just shrug and put up with it. Our Inventor or Hunting Instincts may be motivated to change things, but our Worker Instinct would rather complain than try to fix it. We are the ones who want to have long conversations about things that are wrong with the world that we have no particular control over. In our mind, there is usually someone or something to blame for everything. And we are more motivated to spend our time and energy figuring out who or what is to blame for the problem than we are to try and fix the problem.

Some of us with Worker Instinct even believe the weather is intelligent. If we get a week of nice days in a row, we may be convinced that the weather is now going to turn bad, as if the weather is consciously deciding, "Well, I let them slide with five days of sunshine; now I'm gonna hit them with a bunch of rain just to screw up their weekend plans!"

Worker Instinct is the home of conspiracy theories. We may believe that other people conspire to cause us problems. We are suspicious of governments and politicians. Worker Instinct is generally distrustful of people with authority, but usually will not question it. Workers don't rock the boat.

We can also believe in supernatural explanations for strange events. We believe in UFOs and people visiting us from other planets. We believe in horoscopes and palm reading and fate. We pretty much will believe in many things that we have no control over.

DON'T PUT OFF FOR TOMORROW - Us Workers see that in the long run, very little really matters, and there is no point in sweating things. This causes problems with people with Nurturing, Warrior, and Gathering Instincts. Since Nurturing, Warrior, and Gathering Instincts are all about the immediate issue, they see things very differently. People with these Instincts may judge people with Worker Instinct as lazy and un-caring. What they are really saying, is how us Workers are unlike Warriors, Nurturers, and Gatherers. And yet these people will buy books about not sweating the small stuff, and attend workshops to learn how to take things easier.

I'M FINE - We don't like to talk about health and emotional problems. If we have health issues, we would rather suffer in silence than draw attention to ourselves. We are typically uncomfortable with attention. We will show up sick to work and not even let anyone know. We can be diagnosed with cancer, and will not even tell our children right away because we don't want them to worry. We don't set up regular physical exams and dental appointments. We will wear the same pair of glasses for years and never think to change them. Our children's health will always come before our own. This is something we share with people with Nurturer Instinct, and can be a big reason why those of us with these two Instincts hook up.

CAVEMEN WITHOUT CELL PHONES - There are cavemen who still live in civilized nations and still exhibit primitive behavior motivated by their Instincts. They are the homeless street people we walk by on our way to work. They are the hobos, the bums in the park, and the winos under the bridge. They forage through the concrete jungle looking for food the same way we foraged through the wilderness for 3.5 million years. They seek shelter where they can, and even head south for the winter. They have found the perfect way to exist off the land without working. This is Worker Instinct at it's most extreme end. They see the big picture and have decided that the whole idea of belonging to a society that is based on work and money is not worth the trade off. So they simply opt out of the whole deal. Judge them all you want; at the end of the day there are two gentlemen of leisure: the street people and the very wealthy. It is all the rest of us who are working ourselves to death, so we can have the time to sit around and do nothing.

COMPATIBILITY

MY BETTER HALF - Someone with Worker Instinct can be a stable partner for many others. If you believe the saying that, "Behind every successful man there is a good woman", then that woman probably has Worker Instinct. This can be an excellent base for someone with Inventor or Warrior Instinct to reach out from. Us Workers will always get your back. We are your ultimate buddy, your total friend, and your loyal assistant. We don't ask much, and we are so low maintenance, that we are a joy to be around. Indeed, our relationships are the unspoken envy of many of the other Instincts, who wish they could be this kind of a supportive partner.

THE BALL AND CHAIN - Likewise, since those of us with Worker Instinct tend to take few risks and desire predictability, we can also be too cautious for someone with Inventor or Warrior or Hunting Instincts. We can end up polarizing. We may believe our job is to keep the other person from going out too far, which could become a source of frustration for our partner. An Inventor might start to see us as not being supportive. A Warrior could see us as the enemy. And a Hunter could see us as a dead weight.

PARTNERS

WORKER AND HUNTING - See Hunting and Worker - page 32.

WORKER AND GATHERING - Since most of the people in any country are a combination of Worker and Gathering Instincts, this is probably the most likely pairing that will last until death. Us Workers have the reliability and stability that Gatherers need to feel secure, and us Gatherers can maintain the Worker's world in a constant state of satisfied needs. Both of us have low expectations, and we both will stick to whatever we commit to until we die. Both of us like a little passion, but we will not risk stability even for a brief moment of it. We derive long term satisfaction from being together and being predictable.

WORKER AND WARRIOR - See Warrior and Worker - page 95.

WORKER AND WORKER - Worker and Worker, and Worker and Gathering, are probably the most common pairings since there appears to be more of us with Worker Instinct than any other Instinct. Those of us with Worker Instinct can get along with just about anyone. We have low expectations, and so we are rarely disappointed. We usually marry for life. We fear change, and don't rock boats. Even if some part of our relationship may not be all we had hoped for, we will tough it out and try to make the best of what **is** working. Couples with Worker and Worker Instincts are the fifty percent of all marriages that stay together. Us Workers have the kind of marriages that Nurturers strive for and cannot attain, because of the obsessiveness of the Nurturing Instinct. We live the quiet peaceful life that Inventors wish they could live. We have the supportive companionship in our relationships that Hunters dream of but can't organize. And we intuitively live the kinds of values that our Warrior Instinct lectures about, and tries to force onto other Warriors' lives.

To live with a Worker, is to live a quiet life of predictable and reliable events. We will stay faithful until death, raise healthy happy kids, sleep well at night, and die happy.

WORKER AND INVENTOR - Do opposites attract? If a Worker tries to pair with an Inventor, they may end up polarizing and each feeling like the other one doesn't understand them. A Worker may enjoy the passion of an Inventor, but react with caution to some of their ideas. An Inventor may like the stability a Worker provides, but it could become boring. Part of the relationship will work because they both see the big picture and will pick and choose their fights carefully so as not to jeopardize the relationship. Any conflict might come from feelings of inferiority in the Worker. Or the Inventor may be attracted elsewhere by the promise of a higher level of intellectual stimulation and passion.

WORKER AND ATTRACTION - If one of us with Worker Instinct pairs up with someone with Attraction Instinct, we will probably feel very lucky to have ended up with such a beautiful partner. We will probably bend over backwards to keep them. Our Worker Instinct can make us spoil our partner and love them faithfully until death. But, those of us with Attraction Instinct could easily get bored. Our Instinct is relentless, and we will constantly need reassurance from other people that we are attractive and desirable. This can cause a slow decay in the relationship as the Worker watches their partner flirt with other people. Workers are tenacious though and will hold on and keep trying until the thing slowly spirals down to a passionless state.

WORKER AND MATING - See Mating and Worker - page178.

WORKER AND NURTURING - Worker Instinct may actually be the perfect support system for someone with Nurturing Instinct. Nurturers just want to have a family, and keep everyone healthy and alive for as long as possible. Workers desire predictability, and intuitively exhibit strong family values and respect. There is no problem of either one of us feeling unimportant or not needed. The Nurturer will probably shame the Worker about certain behaviors, (Nurturers shame everyone, it's the natural outcome of constantly judging people, situations, and behavior), but this will feel normal to the Worker. Unlike our Warrior Instinct which cannot stand shame, our Worker Instinct will put up with a lot of abuse for a long time. We shrug it off, pick our fights, and look at the bigger picture. And the bigger picture for both of us will be: a healthy happy family. Sure there's gonna be problems every once in a while, but if most of it works, then why dump it? A partner with Nurturing Instinct could be a good balance, emotionally present, and committed for life. We both have very basic needs, and both are equally equipped to meet each other's expectations. There is no problem of one of us feeling unimportant or not needed. But, our Nurturing Instinct can get on the Worker's nerves if we get too dictatorial about how the Worker should be doing certain things. However, this plays well into our Worker's belief that other people are in control of our lives anyway. Could be a very happy match.

WORKER INSTINCT

IN A NUTSHELL:

It desires predictability and fears change.
It makes us reliable and dependable.
It makes us not want to own the business
or be responsible for other people's security.
It drives us to simply put in our time,
and then go home.

WORKER INSTINCT IS SATISFIED WHEN:

We feel appreciated, or at least noticed.

WHEN IT ISN'T SATISFIED:

It may make us complain,
but we will usually just shrug it off.

GIVEN UNLIMITED RESOURCES:

We would probably not know what to do with ourselves. We may go fishing, play with our kids, plant flowers in our gardens, or lie on the sofa and sleep all afternoon.

INVENTOR INSTINCT

designs and improves the systems that we live and work by. We are the chairmen of the board, the directors, the project managers, and the problem solvers of any group. We see the bigger picture. We remember what happened in the past and contemplate what the future might be. We invent governments and facilitate trade. We organize the Workers and regulate the resources. We negotiate boundaries for nations and debate the laws. We advise the Warriors and consult with the Hunters. We listen to the Nurturers and coordinate efforts toward common goals. We listen to the Workers and act as liaisons between groups. We work with Gatherers and act as moderators over supply and demand. We make decisions based on observation, facts, and proof. This instinctual tendency towards objective observation is a trait we share with Hunters. But, where

Hunting Instinct seeks to find the most efficient way to use current technology, Inventor Instinct seeks to invent a new technology. We Inventors have trouble thinking **inside** the box. The first thing we see is where something can be improved. We Inventors are the visionaries of the group. We often pursue creative lines of work like art, music, writing, design, dance, or architecture. Inventor Instinct invents by not doing things the way other people would do them, like lining up the type on the right hand side of the page.

But, those of us who create and design don't always have Inventor Instinct even though, most of us with Inventor Instinct spend a great deal of time creating and designing. For example, most architects are probably Inventors. But, Inventors can be architects in a larger definition of the word. Inventors may well be the architects of civilized society. The Inventor Instinct could be the reason why we stopped wandering around looking for food and shelter, and started building villages and raising crops. The Inventor Instinct is based on observation. We would have been the ones who realized the seasonal aspect of food that was being gathered, and consequently were the first to think about trying to grow food instead. Finding apples on a tree is great, but if we planted a whole bunch of apple trees, we'd have apples a lot more often, and right in our own back yard too! The whole move from wandering bands of nomadic humans to small farming communities was probably the direct result of our Inventor Instinct. We would have realized it's easier to defend a community if you build some kind of wall around it and stay put. We are the ones who make sure our needs are getting met, and our needs haven't changed in 3.5 million years: shelter, food, water, safety. These things are much easier to guarantee if we stay in one place, chosen for it's resources, and develop it to suit our needs.

We Inventors are the people, some will say, that were born with a gift or talent. Those of us who have Inventor Instinct will tell you there is no gift involved; it is all hard work and dedication. The thing we were born with, that perhaps others were not, is the Inventor Instinct. But, we Inventors can also be the biggest pains in the ass, and the most depressed. Since we usually see the big picture, we often miss the small picture. We want to improve things that are working just fine for the other Instincts. We are constantly thinking, and can be exhausting to be around. Since most of our energy is focused on what could be, but we have to live with what is, we are often discouraged and depressed. We alienate ourselves, and become so preoccupied with lofty thoughts and ideals, that we often overlook the simple pleasures of life. Some of us can't sit around a campfire and roast marshmallows without thinking about ways to make a better marshmallow, a better stick, or a better fire pit. The flip side of this, is that by sitting around and thinking, we Inventors have invented everything from the wheel to the cell phone.

But, we can also be charismatic and charming because we think of things other people don't. We can be very persuasive and win you over with our reasoning. We can rally people around us because we talk convincingly about clear ways to improve situations and we seem to have all the answers. If this Instinct is used to meet other people's needs, it can be a great gift to many people. If used to secure one person's needs it can be a curse. Thomas Jefferson was a person with Inventor Instinct - and so was Adolf Hitler.

PRIMITIVE

Us
Primitive Inventors
would have experimented with new materials for a roof to see if some kinds of leaves blocked more rain that others. We would have scouted out a location to move the tribe to for the winter. We would have been concerned with finding new materials to make clothing out of when the seasons changed. We would have been the ones who discovered that you could mold clay into pots and cook in them. We probably discovered the workability of metal, and pioneered the steps toward being civilized. Because we watched the earth and the way it works, we would have been the ones who would have first thought of planting and harvesting. We would have spent a lot of time sitting and thinking. We may have stared at the stars and watched the way water flowed over rocks.
Consequently, we would have figured out how to navigate the earth and harness its energy.

Inventors
are still inventing.

We still perform the same roles for society, only the technology is different. Where we used to paint with sticks and feathers on the walls of caves, now we use computers. We design the buildings we live in, and the towns, cities, and services. We are usually the brains of the operation and the ones with visions of the future. We are developers and engineers. We are pioneers, artists, and trend-setters. Us modern Inventors are the ones standing in the hardware store with a piece of plumbing in one hand and an electrical device in the other and wondering how to connect them to create the apparatus that we dreamed about last night. There will be a handful of people who will read this sentence and know exactly what I am talking about. The Inventor Instinct is probably the most rare of all.

And it's just as well;
too many chefs can spoil the soup.

TYPICAL BEHAVIORS

Gods and Monsters - Thomas Jefferson was truly a person of vision. His vision was to create a system of government that actually was by, for, and of the people. His noble efforts in trying to establish a functioning democracy were born out of the memory of the tyranny of King George of England. Jefferson sought to establish in people's minds the concepts that all people are created equal and endowed with the same rights. His brilliance as an Inventor, however, is often overlooked because of the stories about how he and George Washington used to smoke pot and have sex with the Negro slaves. His vision of a perfect country was one where all the people had a voice in how the government was run. But African Americans would have to wait quite a while before another president freed them from their slavery, and even longer until they got the right to actually vote. Unfortunately, Thomas had a limited definition of the word "all".

Adolph Hitler had a vision of a perfect world also. His was a world where all the people were actually created equal from the start. His dream was of a master race of superior human beings that were physically and mentally advanced. Of course, this meant that all the inferior human beings had to go. But, being the charismatic Inventor that he was, he had little problem rallying people around him in his cause. Sometimes an

Inventor's motivation is pure, but their own private lives can have problems. And sometimes their motivation is not so pure, and their impact upon society is huge. The cause can vary widely in Inventors. Our history books are full of the effects of their behaviors.

Men Are Pigs - The Warrior and the Hunter can live in a ditch - they often do so as a part of their jobs. They are not concerned at all with the creature comforts of a clean furnished dwelling. They are ready to make do with whatever to get the job done. We Inventors are the ones planting rose bushes and building rock gardens with waterfalls. Our Inventor Instinct is generally concerned with appearance as well as practicality. Our Hunting Instinct may make us buy a brown jacket because it's marked down for clearance and we don't really care what color it is, but our Inventor Instinct will pass on the savings and pay full price for a blue one because we like the color better.

Think Tank - Our Inventor Instinct makes us good choices for heads of research and development. We are excellent managers and group leaders. We keep the focus. We question everything and objectively review the answers. We see trends, directions, rises, and falls. We invent the theories that our Hunting Instinct then tries to prove.

Our inventions can take many forms due to a combinations of Instincts. A combination of Inventor and Warrior Instincts would invent weapons like the atomic bomb. A combo of Inventor and Nurturing

Instincts would invent things to prolong life, like the artificial heart. A combo of Inventor and Worker Instincts would create labor saving devices like the remote control and the drive-thru window. A combo of Inventor and Hunting Instincts can invent more efficient ways to do things like calculators and cell phones. Inventor and Gathering Instincts together might invent ways to get more things faster, like credit cards and convenience stores. Those of us with pure Inventor and very little other Instincts are the philosophers. Invention without a practical application is pure theory.

Genius - Hunters are the researchers of the tribe. They are all about details. They ask, "How, what, where, and when?" Inventors are the thinkers of the tribe; they ask "Why?". If a person with Hunting Instinct also has an exceptionally high Inventor Instinct, you might just have a genius on your hands. Knowledge plus wisdom: this combination of vision, and the dedicated motivation to do the research to make it real, is very rare.

Genius apparently has nothing to do with intelligence. Mozart was a musical genius, but couldn't manage the business end of his life, to the point where he couldn't even pay his rent. Einstein wore mismatched socks. These are classic cases of seeing the big picture and missing the small picture. Excelling in one particular field to the point that you are above and beyond everyone else may not mean that you are smarter than them, it could just mean that you have an excess of Inventor Instinct guiding your Hunting Instinct.

Long Live The King - Most of our leaders become leaders because of our Warrior Instinct. But, those of us with Inventor Instinct are the quiet leaders in any society. While those of us with Warrior Instinct lead by force or threat of force, we Inventors lead by inventing new ways to live our lives. In this way, we quietly change the day-to-day behavior of human beings. The automobile, the telephone and the credit card have changed the way more of us live than any war we've ever fought or medicine we've ever invented. Inventor Instinct creates a very different kind of leader. What's important to us Inventors is the big picture: is there enough to eat and is the water supply protected. Inventor kings rule over peaceful countries inhabited by citizens whose needs are met. They don't start wars to steal resources from others. They invent ways to use the resources they have, to meet the needs of their people. Some of the longest running governments in our history have been those run by monarchs.

The Music "Business" - Inventor Instinct pioneers what's new. Inventor Instinct inspired poor black workers to invent The Blues. It inspired Kentucky coalminers to sing about their lives and create Bluegrass music. Inventor Instinct starts trends. Warrior Instinct turns trends into systems so it can control them. When Hip Hop started it wasn't gangsta, it was happy music born out of funk. Music, like many of the arts, moves in cycles. The Inventor Instinct creates something new, and then the Warrior Instinct tries to reduce it to a formula that can be repeated and generate money. Executives in the music business, who have Warrior Instinct, work hard to find out what people are buying, and then try to make more of that. They insulate themselves from the next new thing by focusing on making a lot of whatever is selling now. Just try to get a demo recording of something new to a major record label and you will bang your head right into our Warrior Instinct. This cycle of Invention and then mass reproduction is kind of like the dog and it's tail: it's often hard to tell who is wagging who.

90% - Some of us with Inventor Instinct get a project 90% completed and then want to start the whole thing over, in a completely different way. The new way will usually be better, but the result is a string of brilliant ideas, none of them seen to completion. The problem is, we lose interest because our minds move so fast, and we want to focus on our newest idea.

We can often be frustrated overachievers, moody and depressed. We function best with limited resources and tight schedules.

Around The Horn - Inventors are unlikely athletes. We aren't interested in competitive sports, or sports that require years of practice to develop skills. Inventors climb Mt. Everest and sail solo around the world. We probably won't even look like the kind of person who would do such a thing. You could sit and talk with us for hours and we may not ever bring it up. What is important to us was the experience of doing it.

Our Warrior will brag about our accomplishments, but our Inventor won't. It's not that we are shy or not proud of our accomplishments, it's just that the reward for us was in doing it. Now it's done, we are sitting there quietly thinking about what we want to try next.

The Sky Is Falling - Our Inventor Instinct is probably responsible for more suicides than any other Instinct. Inventor Instinct sees the bigger picture and sees where things could be improved. This level of awareness can be overwhelming, and leave us feeling powerless to change things we believe need to be changed. We Inventors look at the behavior of people, and the direction it is going, and we realize that there is no way we can stop it. We are the early warning system for the rest of society. We see bad stuff coming long before anyone else. It is our nature. We are the prophets and the messengers. And most of the time, no one listens to what we are saying until long after we are dead. This reality is so obvious to us that it can depress us to the point of taking our own lives. We see too much of the big picture, and we believe the big picture basically sucks.

The vision of Inventors, can often frighten those of us with Warrior Instinct. It can make us feel even more insecure and see people with Inventor Instinct as a threat to our control, and therefore an enemy. But if we see Inventors as being on our side, we may consider them a patriot. Hence the difference between Patrick Henry and Martin Luther King. Our Nurturing Instinct can see Inventors as thinking too negative. In fact,

Nurturers will try to stop Inventors from thinking at all. Nurturers see that it makes Inventors unhappy and will try to counsel them into thinking happy thoughts instead. If our Nurturing Instinct shames our Inventor Instinct, it can make us feel defective, since we are unable to stop it. This can make us want to die even more. This may sound harsh to some of you who read this. But there are those of us who live with these realities every day. To us, it's just normal.

The Brooding Artist - The history of art is full of talented visionaries who end up killing themselves. And not just visual artists, but writers, poets, musicians, comedians, and actors litter the pages of history books with their creativity and their deaths. Virginia Woolf, Mark Rothko, Marilyn Monroe, and John Belushi were all people who showed us the bigger view of ourselves. The Inventor Instinct is the origin of both tragedy and comedy. We Inventors are the clowns, the court jesters, the buffoons. We make you laugh at your own behavior because we see you a little more clearly than you see yourselves. We see your short comings, your down falls, and your weaknesses. How much humor can you think of which focuses on the stupid things we think and do? It makes us all laugh, but it makes the Inventors want to die. And their deaths are called tragic and a great loss of talent. And they had so much promise.

Black holes - We are the quiet kids in the back of the class who spend hours drawing disturbing pictures in our text books. We are the ones who paint our lips and fingernails black and read gothic literature. We get piercings and tattoos at an early age. We make and listen to music about death and despair. We are quick learners and get bored with school because it doesn't challenge us. We are the artists whose paintings and poems are full of passion and life, but all we want to do is sit at the bar and drink. Our Inventor Instinct motivates more of us to become alcoholics and drug addicts than any other Instinct. We Warriors drink because it makes us feel invincible. And then we want to prove it by starting fights. We Workers drink to drown our sorrows. But, we Inventors drink to numb our Instinct, because we see the bigger picture all too clearly. We believe that eventually we all die, and what we do in the meantime is of little or no consequence.

COMPATIBILITY

Worth it? - Those of us with Inventor Instinct are typically loners. We are not really compatible with anyone. We can be vexing and tiresome, enormously entertaining, and damn hard work to have a relationship with. We burn out the affections of the people who love us. We talk ourselves out of our own potential happiness because we see the bigger picture. We understand that the odds are slim on a long term happy partnership, so we won't even bother. We tend to see the brutal reality of relationships and believe that they are often shallow compromises at best, or attempts to avoid loneliness, or orchestrated by the desires of someone's Nurturing Instinct to have families. Inventors are not usually motivated to have children. We see where things need to be improved and we wouldn't feel good about bringing children into the world the way things are.

Having said all of the above, there are ways a relationship with an Inventor can work. We Inventors are motivated by a cause, a mission. We are driven to make things the best they can be. If **you** are our cause, that is, if the relationship is our primary focus, you can have an excellent partner that your friends will be totally jealous of. We Inventors will bend over backwards to make things work. We can be tenacious and committed, thoughtful and supportive. Inventors will periodically reinvent their relationships to stop them from being dull or predictable. This can be scary to Workers, but if you are comfortable with change, there will never be a dull moment. And, if our focus never changes, you can live a long and happy life being spoiled and loved unconditionally.

But, by the same token, if something else in the our life becomes our cause, you and the relationship can get lost in the shuffle.

The Know It All - We Inventors can be insufferable if we think that we know more than you do. Kind of like the guy who writes a book saying that everyone is motivated out of eight primitive Instincts. We Inventors like to debate, even over the smallest things. Like our Warrior Instinct, it's important for Inventor Instinct to be right. And we will probably be "the best" at telling you how you should think and act. The problem is, we will have indisputable facts to back up our argument. Warrior Instinct and Nurturing Instinct are no match for our ability to argue. Since we sees the bigger picture and focus on why something is the way it is, we can confuse both Warriors and Nurturers with reasoning. Warriors and Nurturers have belief systems based on , "This is the way we have always done it, this is the way I am going to continue to do it! You just don't ask why, and that is that!" But, the good news is; that we Inventors are probably going to be smart enough to see when the relationship is threatened by our behavior. This may motivate us to make saving the relationship our cause. Unfortunately, the bad news is; that living with a Inventor may be a continuous roller coaster ride of pissing you off and then making up with you.

Brilliantly Blind - The biggest shortcoming of the Inventor Instinct is that it focuses on such a broad view of the world that it can often overlook smaller issues. Inventors don't remember birthdays, phone numbers, or sometimes just saying "Thank you" when it matters. We can appear to be insensitive and stuck up. Since our focus is on the big picture, we sometimes don't even see what is right in front of us. In much the same way, the Gathering Instinct only sees its actions right here and now, and doesn't focus on the long term effects of its behavior. But this kind of focus is what makes our Gathering and Inventor Instincts effective. The downside is, our partners can often feel that they have little or no importance to us Inventors. We Inventors can become so focused on trying to pioneer a sustainable non-polluting energy source that we forget to eat dinner or call when we are going to be late.

Inventor and Hunting - See Hunting and Inventor - page 32.

Inventor and Gathering - See Gathering and Inventor - page 54.

Inventor and Warrior - The initial attraction can be powerful. To our Warrior Instinct, Inventor Instinct can appear to be Warrior Instinct. We are self-confident, can convince you that we are right, and passionate about what we believe. To someone with Inventor Instinct, a Warrior is easily won over and hence easily manipulated. Inventors can use Warriors to get things done. They work well as the silent partner who suggests things at the right moments, and make the Warrior think it is their idea. We Inventors can be very manipulative. But, the trade-off is, that our partner will be totally spoiled with attention. This couple can work well, as long as the Warrior never figures out that they are being manipulated. Then it could get ugly.

On the other hand, Warriors like to be given direction. Inventors create theories and mastermind policy. Warriors like nothing else than to be given a clear guide of what is right and wrong, and how to handle a certain situation. Alexander The Great, Hannibal, Julius Caesar, and Adolf Hitler all had Inventor Instinct. They dictated to masses of Warriors what they believed was the right way things should be and let the Warriors do the dirty work. And, if Warriors believe they are doing the right thing, they will die for their leader.

Inventor and Worker - See Worker and Inventor - page 119.

Inventor and Inventor - This can be a perfect match. If two creative and thoughtful people partner up, they can have an inspiring, challenging, rewarding, and never boring relationship. This is the kind of relationship where each one feels respected and supported, and the variety of interests keeps them both sharp as tacks into old age - that is, as long as neither partner also has Nurturing Instinct or Warrior Instincts. Then they will constantly be butting heads over philosophies and why each one does certain things certain ways. Might just drive each other nuts. If they have children, they may be driven and brilliant, but potentially unpopular and miserable.

Inventor and Attraction - See Attraction and Inventor - page 163.

Inventor and Mating - See Mating and Inventor - page 178.

Inventor and Nurturing - Our Nurturing Instinct is not really compatible with our Inventor Instinct. Our Nurturing Instinct sees the glass as half full, and we should all be glad that we still have at least a half a glass. Our Gathering Instinct sees the glass as half empty and needing to be refilled. But our Inventor Instinct simply sees a half of a glass. We make no judgment about it. It is simply what it is, a half of a glass. This is a way of seeing the world which confuses Nurturers. Nurturers hear judgements - constantly - even if there are none actually being made. What's more, Nurturers generally hear judgements as negative.There is likely to be constant misunderstanding, resulting in a loss of intimacy.

About 2500 years ago, a Chinese philosopher, Lao Tzu, wrote the *Tao De Ching*. Many people consider this book to be a spiritual guide to how to live a happier life. It is the basis for a religion with millions of members. Lao Tzu was an Inventor who saw the good and bad in human beings. He wrote about the flaws in how we live and interact, and how to improve our lives. But, even

Lao Tzu would have had Nurturers telling him not to focus so much on the bad stuff. We Inventors sees what needs to be improved. We see the downside, the hidden flaws, the stumbling blocks, and pitfalls. Our Nurturing Instinct just wants everything to be all right, and everyone to be happy. Nurturers don't see what Inventors see. And that is, that by studying what is **not** working and why, we can create something that **does** work and improve our lives. Nurturing Instinct is so focused on trying to be positive, that it doesn't see that Nurturers and Inventors are both working towards the same thing, just from different angles.

Nurturers can be attracted to our passion, but they can also be scared of it, and may try to steer us away from following our dreams. We Inventors are risk takers. We are pioneers, and will try things no one else has ever tried. Nurturers like safety and never take risks. This combination can be frustrating to both. Because we Inventors see the big picture in all things, we may see Nurturing Instinct as closed-minded. Nurturers often confuse Inventor Instinct with Warrior Instinct. Consequently they may judge Inventors on a surface level as being stuck up, snotty, always trying to get the upper hand, rude, and judgmental. Our relationship can disintegrate into constant disagreement about the big picture and the small picture, and eventually tear us apart

INVENTOR INSTINCT

in a nutshell:

It makes us thinkers, dreamers, and pioneers.
It makes us focus on the big picture but
we often miss the matter at hand.
It desires wisdom.
It is how we create a better life for ourselves.

Inventor Instinct is satisfied when:

We invent a new technology.
We think of a new theory.
We make something work better.

When it isn't satisfied:

It uses reason and argument.

Given unlimited resources:

We will sit paralyzed by indecision
over what to do.

The Breeding Instincts

The Breeding Instincts are three separate Instincts which combine to insure our continued reproduction as a species: Attraction Instinct, Mating Instinct, and Nurturing Instinct. These Instincts motivate us to look attractive, have sex, reproduce, and nurture our young. Some of us have a high level of Nurturing Instinct without much Attraction Instinct or Mating Instinct. Some of us have a high level of Attraction Instinct and Mating Instinct but little or no Nurturing Instinct. There are many variations in the levels of the three Instincts in a given person, but the behaviors are the same. While there are different biological roles that males and females play in reproduction, both males and females can exhibit the same behaviors relating to these three Instincts.

All three Instincts see the small picture. They focus on the here and now, and try to take care of the issue at hand. Their issues are different, but their focus and behaviors are similar. They make similar partners in relationships, but their attitudes about sex will be very different.

~ Something to remember ~

Human beings do not normally mate for life.

The majority of us act like Serial Monogamists. Over the course of our lives, starting with our first boyfriend or girlfriend, we have a series of relationships. That is, we commit ourselves to being with one partner, and we stay with them for a period of time, and then we move on to another partner, and stay with them for a period of time. It is the period of time which changes - one night, two weeks, three months, 20 years, or whatever. The concept of mating for life is a desire, not an Instinct. If we did instinctively mate for life, we would still be with our first boyfriend or girlfriend, and we wouldn't ever feel attracted to another person.

Also, there appears to be no on/off switch on this Instinct. We humans do not seem to go through a spawning cycle like other animals. We will mate all year long, day or night, and with whoever is available.

"Aren't you cold?"
"I'm freezing!"
"Well why don't you put on a coat?"
"But then no one will be able to see how cute I look in my new top."

Attraction Instinct

Attraction Instinct is concerned with being attractive to the opposite sex. Our idea of what is attractive seems to change constantly. Yet, the Instinct to look attractive stays the same. Those of us with Attraction Instinct think everyone else is judging us on how we look. This can either make us happy, nervous, or annoyed. It all depends on whether or not we are comfortable with the attention.

Primitive

Primitive people adorned themselves with flowers and feathers. They pierced their lips and ears and noses with bones and wood. They scarred their bodies in patterns and invented tattooing. They did dances in costumes to boast and strut and catch the eye of the opposite sex. But, due to the fact that there were considerably fewer people around, they were probably not too picky. Their standards of what was attractive and what wasn't were probably much looser than ours.

modern

Today, we have very clear ideas about what is and isn't attractive. We have beauty pageants and contests that we use to determine and reward each other for being the most attractive. Attraction Instinct is powerful and relentless. The motivation to look attractive, and keep looking attractive, can make some of us with this Instinct miserable, and miserable to be with. We are sometimes called "High Maintenance." We seem to require constant reassurance that we are attractive and wanted. It is a double bind. We may not be able to rationally explain why we are driven to stay attractive, and may be uncomfortable with the attention that it brings us. Since there is no way to stop or reverse the aging process, it is a losing battle. Yet millions of us put faith, time, energy, and money into trying to be as attractive as we can for as long as we can.

typical behaviors

The Dumb Blonde - It's a stereotype. Yes. It's a degrading judgmental assumption based on appearance. Absolutely. This having been said, there are some of us men and women who are so preoccupied with our appearance that we are intellectually and emotionally shallow. With Attraction, and little or no other Instinct to balance it out, we can be beautiful to look at, but hopeless at managing the details of our lives. For us, the smallest decision is almost overwhelming because we are afraid of doing the wrong thing. We would much rather have someone else make the decisions. Since we don't see the bigger picture we can end up doing things last minute and poorly prepared. But we will always look good when we finally show up. Attraction Instinct is probably the most short-sighted of the three Breeding Instincts. Our concern usually stops at how our hair looks, and if our shoes work with our belt or not.

Nip and Tuck - Primitive females would probably have started bearing young as soon as they were biologically developed enough. And they would have been dead long before they had time to sit around and think about the fact that they don't look like they did 20 years ago. Modern females with Attraction

Instinct, may want to continue to attract mates, even though they are no longer bearing young. Remember, this Instinct operates below conscious rational thought. Modern women don't want to have breast implants in order to attract a partner who will knock them up. They may have plastic surgery to feel more attractive. But the motivation to look attractive might be an instinctual left-over of trying to look like they are young and healthy, and ready to breed. Instinct does not pay attention to age or any other factor. There may be no conscious awareness of why they act like they do; all they know is that they are driven to stay attractive and youthful looking.

Some modern women are highly motivated to look like they are not as old as they are, and to look like they have not had children. Face lifts and tummy tucks are generally done to try to regain a youthful appearance. Breast enhancement is performed to create the appearance of younger looking, (ie. firm and uplifted) breasts. Chemical peels of the face are done to create young looking skin. And what about women who haven't had children and want to have their breasts enlarged? Why focus on breast size? Why are some men so attracted to large breasts? Perhaps it's the Attraction Instinct telling them both that a woman with large breasts will be able to provide plenty of milk to ensure the survival of her young? Now we know there is no relationship between breast size and milk production, but our primitive ancestors may have believed it. Remember, there were no milk-giving cows around for most of western history. And so some women and some men may believe the ideal look for a woman is young with large breasts, or to put it in Attraction language, to be at the age where she is ready to start breeding and with plenty of milk to suckle her young. Even The Bible gets into the act : "We have a little sister, and she hath no breasts: what shall we do for our sister in the day when she shall be spoken for?" (Song Of Solomon 8:8)

Makeup - The history of makeup can be traced directly to prostitution. Lips were painted red to mimic the red swollen lips of the vulva flushed with excitement. Rouge was applied to the cheeks to create the illusion of youthful blushing or orgasm. Dark lines are drawn around the eyes to make them look larger, more like an innocent child, to catch the male's attention. Prostitutes don't want to get pregnant. And the males who have sex with prostitutes are not trying to get them pregnant. Prostitutes are simply presenting themselves as willing partners to have sex with. This is highly attractive to a male with Mating Instinct.

Soon, other women saw the effect that this makeup had on men and began to use it to conceal their age and appear more attractive to a potential mate. Modern makeup has now expanded into the production of specific products which claim to slow or even reverse the aging process. All of this is to create the illusion that the woman is younger than she really is, even though once the male gets close enough, he will see that she is not actually as young as she appeared to be at first. However, a male with strong Mating Instinct is probably not likely to turn down any opportunity to have sex.

Anthropologists tell us that we live three to four times longer than primitive people lived. With no genetic blueprint past the age of 30 for over 3 million years, perhaps this behavior is simply motivated by Instinct to try to look 25 years old forever.

The Object Of My Desire - Some of us with Attraction Instinct can be confusing to be around. We behave sometimes in contradictory ways. We diet, work out, and shop for clothes that accentuate certain parts of our bodies and diminish others. We wear makeup and even have surgery to improve the looks of our various body parts. Yet, when someone notices that part of us we may get angry for being objectified. This may make sense in our minds but it is a mixed message to you. Why get breast implants and wear a low cut blouse if you don't want men to look at your breasts? Which came first: the chicken or the egg? Which came first: the skinny model with big breasts, or the men who like skinny models with big breasts?

Some women with Attraction Instinct modify their bodies to make themselves feel better about the way they look. The problem is, males with Mating Instinct are going to be attracted to them specifically **because** of the way they look. But, just to confuse matters more, some people with Attraction Instinct actually **want** you to objectify them. They like the attention because it makes them feel good. It can be difficult figuring out whether they actually want the attention or not. A woman with extremely high levels of Attraction Instinct might be attracted to modeling, stripping or prostitution. To be admired and paid money, because you are so attractive and desirable, would be very satisfying to this Instinct.

People with Attraction Instinct are often called teases. They act like they want to mate, but actually just want to feel desirable and desired. Mixed messages are created by mixed Instincts trying to get their needs met. If a woman with Attraction Instinct teases or flirts with a man with high level of Mating Instinct, there can be trouble. If that man has a combination of high Warrior and Mating Instinct, it can lead to rape. Warriors get their needs met through force or threat of force.

Some things which people with Attraction Instinct use to objectify themselves are: tight clothing, see-through clothing, removable pads, push-up bras, hose, high heels, and articles of clothing that barely cover their genitals and breasts.

Samson and Delilah - Some of us believe that it's not OK to go bald, because we don't look as attractive to the opposite sex. The Attraction Instinct in us Modern males may be motivating us to appear as young and as healthy as we can so we can attract a potential mate. Otherwise, why would we wear a cap of artificial hair on our head, comb over what we have from one side to the other, or spend a bunch of money on medication and surgery? Also, since we didn't used to live much beyond the age of 25, and most people don't show pattern baldness until well into their twenties, we may see it as a negative thing, simply because for 3.5 million years we never saw it.

"Does this hat make my butt look too big?"

"Venus Of Willendorf"
c. 24,000 - 22,000 BCE

A Flat Stomach In Only 10 Days! - The artifacts that we have of primitive people clearly show their tastes. From the first days of civilization up until recent times, artists have created images of desirable women. The *Venus of Willendorf*, one of the earliest of these images, was a large woman with big round stomach and big breasts. Many believe this was some kind of Goddess figure and the statue was carved for religious purposes. It's also just as likely that it is a portrait of someone, or a piece of early pornography carved to be used as an aid to masturbation. Today, there are magazines, videos and Web sites full of pictures of large naked women. It appears some men's taste in women has not changed in a very long time.

The 17th, 18th and 19th centuries in Europe are considered by many art historians to be the peak of visual art in civilized western society. During this time, artists depicted what they believed to be the ideals of beauty. The images of women they painted and sculpted would be considered fat by today's fashion standards. Venus, the Goddess of love, the most beautiful woman they could imagine, is usually depicted in paintings and sculpture at about a modern size 14-16.

Not too long ago women used to wear bustles. These were bunches of fabric gathered over their buttocks to exaggerate its size and draw attention to it. It was considered fashionable. Today, some women try to diminish the size of their buttocks. Some women show off their booty. Some men like a little junk in the trunk, and some men like tiny butts. What is good and bad can change sides with Attraction Instinct. This makes these people a little more open minded than those with Warrior Instinct, for whom wrong and right are, and will always be, wrong and right. This difference is also the source of much disagreement between Warrior and Attraction Instincts.

Less than fifty years ago, the ideal measurements were 36-24-36. Now the standard is towards zero body fat. A few years from now that standard may change again. But, no matter what the fashion trend may be, what remains constant is that some modern women are highly motivated by Attraction Instinct to do whatever they can to stay attractive to potential mates. The diet and nutritional supplement business makes millions of dollars from feeding this Instinct. The fashion industry likewise does their part. More size 12 dresses are sold than any other size, but the average size model is under a size 6 and the average size mannequin is a size 2. More than half of the women in the United States are larger than a size 12.

Working Out - One lifetime ago, weight lifting and body building were pretty much the activities of men only. Recently more and more women have taken it up. For some men and women, Attraction Instinct might motivate them to try to present a strong lean muscular appearance to potential mates. In essence: to look like they are 25 and in the prime shape of their lives. And we have invented artificial ways to make our bodies continue to look lean.

Exercise machines, health clubs, and nutrition supplements are all part of a huge industry which feeds the motivation to look like we are physically fit. And the word "look" is the key, because the best developed bodies in the world are judged by how they look, not by feats of strength. In body building competitions, they focus on the appearance of the muscles, the definition, and the overall picture. The people who compete in the World's Strongest Man and Woman contests, have too much body fat to be admitted into a body building contest. It is an entirely different set of criteria that they are being judged by.

As far as our diet goes, those of us with Attraction Instinct usually refer to our Warrior Instinct and Nurturing Instinct to get our information about what is healthy and what isn't. Our Warrior Instinct used to believe that fat was the enemy, and now it believes that it's carbohydrates. Next year it may change again. Our Nurturing Instinct tends to get lost and confused by new and conflicting information. Don't eat red meat because it's bad for your heart? Or do eat it because it's a high protein, low-carb diet? Eventually we may all come around to realize that the only diet that is really good for us is the one on page 43.

There are also some of us who don't feel motivated at all to change the physical appearance of our bodies. Perhaps this separation indicates those of us with Attraction Instinct and those of us without it.

People with Attraction Instinct work out to **look** like they are in prime physical shape. People with Nurturing Instinct work out to actually **stay** in prime physical shape so they can live a healthy life for as long as they possibly can. And Warriors work out to stay physically strong to fight other Warriors. As it is with all these Instincts, some **behaviors** may be the same, but the difference is in their **motivation**. **Why** is the key. Why are we doing this behavior? If you ask why, you can identify the motivating Instinct.

Middle Age Crazy - Attraction Instinct and Mating Instinct drive us to be attracted to, and want to mate with, the prime physical members of our species. Some men, when they reach middle age, are attracted to women half their age. And some women, when they reach middle age, seem to be attracted to young muscular men who are in their sexual prime. Today, we live an average of about 50 years more than when we were

Primitive people. If we have no blueprint that guides us past the first 25 years, then is it possible that we are just trying to relive that first 25 all over again? For the first 3.5 million years, 25 year olds were the top, the best, the ultimate. And then we died. We never saw 40 year old people. Of course we want to have sex with 25 year olds. Of course we want to be 25 again. It's all we knew for a very long time.

partners

Attraction and Hunting - See Hunting and Attraction - page 33.

Attraction and Gathering - See Gathering and Attraction - page 54.

Attraction and Warrior - We Warriors often aggressively pursue beautiful partners. An attractive spouse can make us look better. We also have that possession booger inside us that likes to have things that other people don't, and rub it in their faces. Our partners can get spoiled and showered with gifts. Both of our Instincts are concerned about how we look to other people, and working out together can bond us. A desire for expensive cars, clothes, and toys can keep this relationship happy. But, our Warrior Instinct will always feel territorial and get angry about other people being attracted to our partner. Jealousy can often turn these relationships abusive.

Attraction and Worker - See Worker and Attraction - page 120.

Attraction and Inventor - Polar opposites, but we may be magnetically attracted all the same. Our Inventor Instinct makes us admirers of great beauty, and people with Attraction Instinct are usually fascinated with people who think about deep subjects, and have big goals. Beyond the initial attraction however, there will be little to hold us two together. We Inventors focus on the big picture and those of us with Attraction Instinct tend to see a narrow view. Both of us can feel judged by the other person as being defective in some way, and the conflict could tear us apart. If both have another Instinct that balances us out, then we may be able to understand each other. But it will probably take constant work to keep perspective on track in the relationship. The one with Attraction Instinct could be seen by the Inventor as shallow and dull. Inventors need intellectual stimulation, and value wisdom.

Attraction and Attraction - The perfect couple! The perfect couple? You would think so, but is there enough between them to support a long-term relationship? And would they become jealous of each other? And what happens when they visibly start to age? Again, probably not an issue when we only lived to be 25, but 50? 60? 70?

Attraction and Mating - See Mating and Attraction - page 178.

Attraction and Nurturing - Another perfect couple? The one with Attraction Instinct gets the care, support, and validation they need to follow their Instinct to stay attractive, and the one with Nurturing Instinct gets to feel useful, needed, and supportive. But can it last? This couple stands the best chance of working, because of all 3 of the Breeding Instincts, the Nurturing Instinct will do whatever it takes to make things work.

Mating Instinct

Mating Instinct is all about having sex. It's the hook-up, it's doing the dirty, it's gettin slippery, it's dibbling and dabbling and doing the snoofer snoo, it's makin bacon, it's slammin ham, it's gettin it on, and bumpin uglies, and, it's the reason why there are 6 billion of us hanging around the camp. Those of us with Mating Instinct are fairly easy to spot. We usually look apprehensive, nervous and anxious. Our eyes dart about the room when we talk to you. Our heads turn whenever someone enters or leaves the room. We are constantly on the lookout for a sex partner. Those of us with Mating Instinct desire frequent sex, preferably with a continuous supply of interesting and attractive partners.

Primitive

From puberty, until they died, Primitive people probably mated whenever they could, and with whoever was available. Primitive humans had an average life span of 20-25 years. If you look at the way hormones start raging in teenagers today, you can well imagine that there was rampant snooferage going on at all times. Add to this, the fact that the infant mortality rate was high, and few people survived to adulthood. The instinctual motivation to have sex any time, anywhere, and with any one, is one of the main reasons why the human race has survived.

Modern

Every weekend at our local bar or nightclub we can see our Mating Instinct in full swing. Looking for a sex partner is the mission of our Instinct, and, it generates a lot of business and industry. It's used to sell beer, cars, clothes, and just about everything else that can be tied to having a happy sex life. Because we all want a happy sex life, right? Or at least all of us with Mating Instinct do.

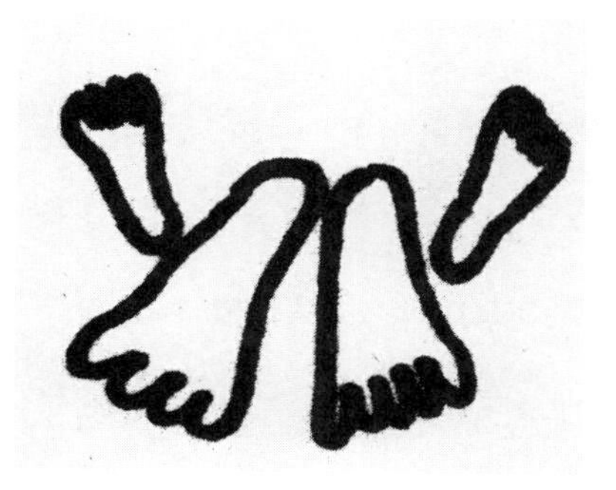

Typical Behaviors

Scanning - One telltale sign that we have Mating Instinct, is that we scan just about every member of the opposite sex we encounter. This can be obvious or subtle, and it takes a fraction of a second. When we scan another person, we do a once over glance from head to toe, and then rate that person on their mating potential. If you could slow the process down and get us to tell you what we just thought, it might go something like this:

Nice breasts
Teeth are crooked
Pretty eyes
Don't like the lipstick
What's up with that hair?

Or

Nice buns
A little fat in the gut
Nice smile
Wish he'd shave
Did his momma dress him?

Bottom line is: Would we have sex with this person? If we have Mating Instinct the answer is usually going to be yes. Remember, for 3.5 million years there were not many humans on the earth. If we did encounter someone we hadn't seen before, it was a potential new partner. Our Instinct is still based on survival of the species, even though there are over six billion of us alive now. This is instinctual behavior; it is highly unlikely we can grow out of it, or force ourselves to stop it, in much the same way, as it is highly unlikely that we will grow out of, or force ourselves to stop nurturing children or gathering food.

One Track Mind - "**All** men think about sex **all** the time." How often do you hear this? It is not true. What is true is: Men with Mating Instinct think about having sex constantly - about as frequently as women with Nurturing Instinct think about the safety and health of their children. It is instinctual behavior. If you go to bars and nightclubs you will encounter men and women who are looking for sex. That is the primary reason bars and nightclubs exist. What you will not encounter, is the men and women who are happily at home with their spouses and kids. And, single people who don't have Mating Instinct, don't hang out in bars. So, if you are looking for a peach, why are you wandering around in an apple orchard?

Why People Cheat - Those of us with Mating Instinct are driven by our Instinct to have sex outside of our primary relationship. For us it is all about seven and a half minutes of hot sweaty nakedness, and hopefully an orgasm.

Those of us with Inventor Instinct might be tempted to have sex outside of our primary relationship if we encounter someone with an interesting shaped body or different ethnic background. We enjoy new experiences, and like to sample the whole menu.

We Nurturers might have sex outside of our primary relationships if the intimacy level has faded for us. We crave intimacy above all else. Just to see the adoring look in our lover's eyes, is almost more important than the physical act.

Our Hunting Instinct likes the thrill of the chase, the stalking and bagging of the prey. We see it as a conquest, and regard the person we cheat with as a trophy, and will often brag about what a catch they were.

Some of us are forced into having sex by Warriors with high levels of Mating Instinct. This is rape. We may be forced with either physical force, or threats and emotional pressure.

Us Gatherers just can't get enough of anything, including love. To be wanted and held, even for only seven and a half minutes while our partner has an orgasm. For those of us with Gathering Instinct, it's about feeling emotionally full. And this will temporarily fill us.

We Workers don't screw around. We mate for life, and would not risk the security of our relationship for anything.

That Funky Stuff - Infatuation is a powerful thing. And it usually leads to having sex. It starts in our teens, with that first incredible rush of emotion and passion. And pretty soon we are out in the back seat of a car dibbling around in each others pants, not exactly sure what we are going to touch. As we get a little bit older, we know exactly what we want to do with our infatuation. During the first few months of a new relationship we usually spend getting slippery every chance we get. Could infatuation be the way our Mating Instinct is still trying to insure the survival of the race? The seemingly random way we are attracted to someone, and the intense bonding and desire to have sex are fairly universal among us human beings. We can't get enough of that funky stuff.

Forcing The Issue - Back in the day, when we were still primitives, and before the invention of birth control, there was a good chance the female would be pregnant after 2 or 3 months of steady sex. At this point she would have different needs and probably spend a lot of time hanging with her female friends around the camp. The male might be needed elsewhere to continue hunting and defending. Perhaps another female who saw that he was able to get the first female pregnant now wants to mate with him. This may have been all well and good for primitive people, but then we came up with the idea that we are supposed to stay together forever with the person we have children with. Are we contributing to the number of failed marriages by trying to force this value onto the Mating Instinct?

After a few months infatuation usually winds down and frequency of sex becomes less. Then there comes a time where each partner looks at the other one, and asks themselves if they want to keep the relationship. Some Modern women still seem to be attracted to the healthy muscular Warrior type, and want to be romanced by them and feel the passion. But once they are married and pregnant they need him to turn into a good provider. Their needs change, and they want someone around them who acts supportive and nurturing, like their female companions. But now they are supposed to stay with the guy who gets them pregnant, so they try to change the man they **have**, into the man they **need**. But how effectively can an Instinct be changed? Could the rising number of divorces be due to the fact that we are trying to force our Instincts to be different?

Green Light, Red Light – Some Modern women are so closely tuned to their Instinct that they actually say, "My biological clock is ticking," indicating that they are very aware of their window of opportunity to bear young. In western society, this is probably the best time in the history of human beings to be a mother.

With the advances made in modern medicine, the odds on survival for children have been greatly improved. Females can pick and choose when and why they want to have children, and with who. Education and health care are at an all-time high.

An interesting development is that today a woman's Instinct to bear young has entered political and legal arenas. A female's desire to bear young is considered so precious it is called a right. There is much discussion over a female's reproductive rights. There are organizations, lobbyists, activists, charities, college classes, hot lines, volunteer agencies, religious groups, and laws all supporting and protecting this basic Instinct. There are religions which openly encourage women to bear as many children as they want and forbid birth control. The effect of all this is that females have the green light to follow their Instinct, and much encouragement and assistance in doing so.

Likewise, this is probably the worst time in the history of human beings for men with Mating Instinct. Modern men in western society are only supposed to have sex with, and children with, the woman they are married to. They are not supposed to have sex outside of this relationship, and many women don't even like them thinking about having sex with other women. And, unlike the females, there are laws **against** a man`s reproductive Instincts. It is the subject of ridicule and shame. There are organizations, lobbyists, activists, college classes, hot lines, volunteer agencies, and religious groups which strongly oppose this behavior, and seek to punish men for following their Mating Instinct.

Shooting Blanks - Getting a vasectomy or having our tubes tied is often done to prevent the possibility of getting pregnant. But for some of us men there is a pride factor in being able to father children. A vasectomy may make us feel like less of a man, and so we joke about it. But under the joke, and the reason for the joke, is the often a different feeling. If we weren't going to stay with the same women until we died, we may want to father children with someone else. Often, we men who get divorces and remarry, get our vasectomies reversed. And for some women, if money and help was not a factor, they would keep on having children.

Pornography - Many of us modern men turn to prostitution and pornography for an an outlet for our Mating Instinct. And there are pros and cons for these behaviors. Prostitution and pornography can erode the intimacy level between a husband and wife, and contribute to a general degrading view of women. But, it also satisfies some women with high Mating Instinct, and provides them with an income. It is usually a poor substitute for having sex with someone we love, and it is against the law in some places. Prostitution conflicts with some religious doctrine, but it has been around longer than most religions, and the concept of marriage. Prostitution is likely to continue as long as there are human beings with Mating Instinct.

Pornographic magazines, videos, internet porn, peep shows, and strip clubs are all generated out of trying to satisfy this Instinct. Strip clubs are the closest we married men can get to having sex with another woman without feeling like we have committed adultery. All these things also erode the respect men have for women. They are also against the law in some places, and conflict with some religious doctrine. Pornography can also become addictive. But, for some of us men, it is the only sex we have. And, at this point in time, masturbating to pornography may well be the safest sex there is on the planet.

A Little Romance Please - Where pornography for men can be about visual stimulation, pornography for women can be about romance. Modern women like romance novels, romantic movies, and romantic music. They plan romantic weekends at bed and breakfasts, they decorate their homes with romantic Victorian decor, and they keep scrap books of pressed flowers from their first dances. It is Instinct driving them to keep recreating the feelings of the initial infatuation period.

The huge popularity of soap operas supports this Instinct. These are not well-written dramatic works, and they take three episodes to complete a whole sentence. These are melodramatic stories, full of emotion and passion, and the romantic plot lines keep going forever. And this is exactly what some females with Mating Instinct desire in their own lives - passion and emotion, and a romantic plot line that goes on forever. And so, they keep tuning in and watching, because if you can't have it in your own life, then the next best thing is to have it in a TV show, or a movie or a novel, and project yourself into the situation. This kind of fantasy projection is universal among both men and women with Mating Instinct. Some women like to project themselves into passionate relationships, the same way that some men like to project themselves into passionate sexual experiences. The need this fills in some women, is the same need that gets filled in men who masturbate looking at images of naked women.

My Cell Mate Thinks I'm Sexy - When a man is in prison, common sense would seem to dictate, "Well, I'm not around any females, so I won't be having sex for the time I am in here". But there is a percentage of us men who seem to be compelled to have sex regardless. We are not gay men who are enjoying making love to another man. We are tough Warrior-type heterosexual hombres. The only human we can have sex with is another man. So we pretend that he is a woman. Why on earth would a heterosexual man do this? Unless, we are being motivated by a Mating Instinct that drives us to have sex, regardless of our situation?

And, if we also have Warrior Instinct as well as Mating Instinct, we may force others to meet our needs, and we may strongly believe we **have** to have sex. This combination of Instincts could also explain other forms of rape.

Taking Vows - What about priests who force young boys to be sexual with them? In some religions, men who join the priesthood, take vows that they aren't going to be sexual with anyone for the rest of their lives.

Unfortunately, if they have Mating Instinct, they could have problems. These are instinctually driven behaviors. They are not controllable by reason or logic, or will power or belief in supernatural beings. This person is going to find a way to have sex, the same way someone with Nurturing Instinct will find ways to keep their children alive and healthy. Instinct is Instinct. It does not discriminate between wrong and right, good and bad, healthy or unhealthy, legal or illegal, or moral or immoral.

Boys Will Be Boys - In a day and time when evangelical Christian groups, and militant heterosexuals, are overtly opposed to gay people, what could possibly motivate someone to choose a lifestyle which might get them killed? - unless they have no choice, and they are being motivated by a primitive Instinct. There is not only one reason why some people are attracted to their same sex; there are several motivators, from childhood sexual trauma to simple curiosity. But could one of these motivators be a chromosomal arrangement which fosters certain Instincts? Since men and women respond differently to Breeding Instincts, are these primitive Instincts part of some genetic code for each sex in the chromosomes? What if a male Mating Instinct showed up in a female? The male side of the Instinct motivates them to want to have sex with women. Could this explain some lesbian behavior? And could some gay men be males that have female Nurturing Instincts motivating their behavior? Male transvestite hookers dress as women and parade the street waiting for men to think they are sexy women, and have sex with them. That may be the extreme end of it, but what is motivating their behavior? Could it be female Mating Instinct motivating them to look like women, and have sex with men?

There are many transgendered people who actually say, "I feel like a man living in a woman's body" or vice versa. What else could possibly motivate someone to undergo surgery to change their sex unless they are being motivated by a strong Instinct? Voluntarily having your sex organs removed is not something done lightly or experimentally. These people are being motivated by something bigger and more powerful than logic or rational thought.

Partners

Mating and Hunting - See Hunting and Mating - page 33.

Mating and Gathering - See Gathering and Mating - page 54.

Mating and Warrior - See Warrior and Mating - page 97.

Mating and Worker - We Workers are no dummies. We may enjoy the sex for a while, but we will quickly see the shallowness of the relationship. Workers like predictability and stability, neither of which, the one with Mating Instinct can offer us. Workers look at the big picture, and Mating Instinct is all about immediate gratification.

Mating and Inventor - Another good combination for a long term affair. These two Instincts could not live together, both of us would get bored. But to see each other on a regular predictable basis would fit our needs perfectly. Our Mating Instinct desires an interesting and willing partner, and we Inventors desire a little pleasant diversion from our minds every once in a while. However, if the one with Mating Instinct screws around outside the relationship, the Inventor may take it personally and can become depressed. Us Inventors are passionate people. It's not uncommon for us to cut off our own ear, or kill ourselves because of despair and rejection.

Mating and Attraction - Another almost perfect couple. The one with Attraction Instinct gets to be admired and desired because of how they look, and the one with Mating Instinct gets to have sex with an attractive partner. But can it last long-term?

Mating and Mating - Sex all day, sex all night, sex, sex, sex, yippie! It can last for a while, but each partner will eventually get hungry for something else and go back to the buffet table to find a new dish. Instincts cannot be satisfied.

Mating and Nurturing - Wham, bam, thank you Ma'am. The initial attraction for the Nurturer may have been that they were looking for a partner to have kids with. What they end up with is just a lot of sex. Those of us with Mating Instinct will pretty much say anything to get a little action. Eventually, when the "M" word comes up, we will run for the hills. Those of us with Nurturing Instinct are easily misled and taken advantage of by those of us with Mating Instinct. These hook ups are usually short term relationships with a big potential for hurt feelings. Also, since we Nurturers are generally attracted to the same type of person, we will repeatedly have the same experiences. But we will not see our part in the pattern, we will simply lament that all the people we meet are always the same, and just want to use us and dump us. Our Nurturing Instinct, being focused on the issue at hand and not seeing the bigger picture, is blind to the common denominator in all of our relationships: us.

Nurturing Instinct

NURTURING INSTINCT focuses on keeping everyone healthy and alive for as long as possible. Those of us with Nurturing Instinct are constantly looking out for potentially threatening situations. Constant vigilance is how our Instinct stays effective. Consequently, we tend to view people, behavior, and situations as either good or bad. We Nurturers bond with other Nurturers by agreeing with each other`s judgements. We just want to help any way we can. We are happy when we feel needed and useful.

Nurturing Instinct drives us to care for people. Specifically, we are motivated to take care of our own children. But this can also extend to taking care of the elderly, our friends, our neighbors, our fellow countrymen, our race, and our species. Some of us Nurturers who are too young or too old, or otherwise unable to have children, will get pets as a substitute. We need someone or something to take care of. A dog that is totally loyal, unconditionally affectionate, and depends on us for its basic needs, can easily provide us with daily satisfaction for our Instinct.

Primitive

Before the invention of hospitals, drugs, and preventive care, Nurturing Instinct would have been very simple: keep the children healthy, and try to cure people when they get sick. Healing the sick was probably done through natural medicine with available plants, and faith healing, through belief in some form of religion. Infant mortality was considerably higher. We lived much shorter lives, and died from "natural causes" such as viruses, infections, and animal attacks. However, since we only lived about 25 years, we were probably fairly healthy. People with Nurturing Instinct who are now reading this will be saying to themselves, "Things are so much better now that we live longer, and the "natural causes" that used to kill us are easily avoidable." Nurturing has come a long way.

MODERN

Our Nurturing Instinct is still concerned with improving human beings. Those of us with this Instinct are motivated to find ways to live longer, happier, and healthier lives. Now that we are living longer than 25 years, we need to add maintenance to our lives to stay healthy. In primitive times we didn't live long enough to see cancer, strokes, or heart disease. Most of the major diseases that kill us, rarely happen before the age of 25. Back then, we were all beautiful healthy 25-year-olds in our prime, eating whatever we wanted, and then we died. Things are very different today. After 25, we have to supplement our diet, set up support systems to combat the effects of aging , and often financially bind ourselves to medical institutions to delay dying. We are driven to try to stay healthy 25-year-olds forever.

Nurturing Instinct is sometimes called the "Mothering Instinct". Although it is probably more common among women, this title is a little misleading because many of us men display the behaviors associated with this Instinct as well. We usually put the needs of other people, especially children, before our own needs. Our primary focus is on making decisions between good and bad. In primitive times it might have been, "Is this a good mushroom for my child to eat, or will it kill her?" Today, although the circumstances are different, the behavior is the same.

Those of us with Nurturing Instinct are generally cheerful and optimistic. We look for the good in everyone and everything. We don't like unpleasant situations, and will seek to find the silver lining and positive angle. Of course we do. It's our Instinct. In fact, those of us with Nurturing Instinct will probably not like this book, because it talks about the negative side of these Instincts. We Nurturers tend to believe the most important thing is to be positive. We want to read books that tell us how to fix things, and what is going to make our lives better. We don't want to read sad depressing facts. This is why our Instinct is short sighted, because it makes us skip over anything negative, and miss opportunities to learn and grow. Our first reaction is to avoid the bad and look for the good. All three of our Breeding Instincts usually believe in myths over facts. It is a trait we share with Warriors and Gatherers. We just want to believe everyone is going to be all right. This is the essence of faith. It may not be factual or realistic, but it's what we believe, because it's what works for our Instincts.

Much of our life is face-value judgment. We judge everything. We have to. It's our job. Judgment is such a constant in our lives that we really just don't need any more, thank you very much. We will avoid anything bad at all costs, even if we have to pretend to be happy when we aren't. Where it's more important for a Warrior to be right than to be happy, it's more important for us Nurturers to be happy than be honest. We just want all of us to be happy. Is that so bad? Because being happy, we believe, will make our lives easier, longer, and healthier.

Typical Behaviors

No Time To Breathe - We are the parent in the mini van who is usually rushing between our son's hockey game and our daughter's swimming lesson. Our partner is picking up the youngest from day care and stopping to grab some dinner for everyone on their way home. Our Nurturing Instinct can motivate us to sacrifice our own needs for the needs of our kids. We will openly lament about our hectic schedule but seem to be unable to fix it. Our days seem to "get away from us". Our Nurturing Instinct focuses on the immediate issue, the task at hand. We do not see the big picture of our day and schedule our time in an efficient manner, like a Hunter would. We are not intuitively efficient about planning time and resources, and instead spend our whole day running from one thing to the next. But, since our motivation is to give our children every opportunity that we had or didn't have, we will rush on, shrugging our shoulders and saying, "Oh well, I guess that's just the way things are."

This creates one of the biggest conflicts that we Nurturers have to deal with. We are often internally torn between what we "should" do and what we "want" to do. We are instinctually compelled to put the needs of our children before our own needs.And we will do it time after time. But some of us will not feel internally satisfied doing this. There will be a nagging sense that we are getting cheated out of things we want to do for ourselves. And then we may shame ourselves about feeling that way, or feel guilty about having desires of our own, and often end up angry at ourselves, our children, society, religion, or whatever we can point our frustration towards.

The Goldfish Principle - Those of us with Nurturing Instinct desire bigger houses, newer cars, better schools for our kids, nicer clothes, and better quality food. We just want to give our young the best we can. This desire will drive us to stay on the brink of being one paycheck away from the street. We may go so far as to risk the security we already have, just to possibly have something better. We will push ourselves to work more, work harder, and stress ourselves out. We can make ourselves miserable in the attempt to have the new thing, the better thing, the bigger thing. And, since this is instinctual behavior, we will believe that it is what we are supposed to do, and will see ourselves as powerless over it: "I just have to find a way to afford to get my child into a private school!"

Spoil Me! - Those of us with Nurturing Instinct will bend over backwards to meet the needs of others, especially our kids. Consequently, we may expect the same behavior to be returned. Since we spend most of our energy and time caring for someone else, we can feel the need for a little time and attention spent on us. We may seek ways to feel pampered and special. We might enjoy going to a spa for the day and getting massages. In fact, spas may well be invented by, designed by, staffed by and frequented by Nurturers. A massage is the perfect thing for someone with Nurturing Instinct. This is all we ask for: is someone to spend some time and effort spoiling us for a while. After all, we deserve it, don't we?

It's difficult for us Nurturers to relax. It's not a normal state for us. Our Instinct keeps us alert to potential danger. Workers, on the other hand, don't go to spas or get massages. It's difficult for them to get stressed.

Teach The Children - A female with Nurturing Instinct can undergo a personality change when she starts to have children. She becomes focused on the health, safety, and education of her children. Children learn by repetition. Every time they play outside, they need to be reminded to wipe their feet and wash their hands so that it becomes second nature to them. Some of these Nurturers however, don't have an on/off switch on this behavior, and may turn to their adult partner and talk to him in the same way. A person with Nurturing Instinct often initiates conversation from the belief that you need help.

This can cause some problems with her partner. When they first started dating, they probably conversed like two adults. Now, she is talking down to him. This same woman is reminding him to wipe his feet every time he walks in the door. Gradually she has changed, and now she talks to him as though he was also a child. Her Instinct is in control. And, her partner may react angrily: " Hey, who are you talking to? I'm not a child, I'm a full grown adult just like you. Do you think I have suddenly gone stupid and don't remember to wipe my feet any more?" What has happened to that great person he married who used to think he was at least as smart as her? She has become a mother, and her Nurturing Instinct is in charge. And when he tells her he feels insulted by her behavior, she is likely to be confused and angry.

Our Nurturing Instinct acts on the belief that since we are motivated to help you, that you would welcome our help. After all, we are only trying to be helpful. We are unlikely to see a difference between **motivation** and **behavior**. Or we may believe that since our **motivation** is to try to be helpful, that any **behavior** we do is acceptable. Most of the time, however, we won't have any awareness of what we are actually doing. Our Nurturing Instinct focuses on the moment, the immediate issue at hand. This is instinctual behavior, not rational behavior. We would not rationally or consciously choose to do something that would destroy the intimacy between ourselves and our partner. And, it is not rational to insult someone we are trying to help.

This kind of behavior was probably not a problem for the first 3.5 million years because we only lived to be 20-25 years old and we had not invented marriage. But trying to stay with the same person for 60 or 70 years is a different matter. When this behavior starts, several things can happen. The male can feel insulted and angry and not know why. The female can feel misunderstood and shunned and not know why. The male can leave to find another female who talks to him as an adult ("My wife doesn't understand me"). They can both enter couples counseling and try to learn to communicate better. This means they will try to learn to think like the other person, and work at meeting the needs of their Instincts. The problem here is, Instincts don't naturally think alike. And their first thought will usually be towards their **own** Instinct, not the other person's. You will have little luck trying to counsel someone out of an Instinct. This behavior can kill the intimacy level between a man and a woman. But this behavior has also been responsible for keeping children alive for 3.5 million years.

For The Kids - Those of us with Nurturing Instinct will probably do whatever we can for our children. And this could mean staying in a relationship that has no intimacy. The message we are giving to our children is that no matter how bad your relationship is, you stay with it. On a common sense level, we would probably not want our child to grow up believing this, and getting into a similar situation. Why on earth would we want to model a dysfunctional, angry, uncommunicative, abusive, or intimacy-deprived relationship to our children? Nurturing Instinct can override common sense and make us stay in a dead relationship regardless.

But these relationships that fall apart can be doomed from the start by instinctual behavior. Some modern women with Nurturing Instinct don't date, they are actually interviewing men to be the father of their children. This is probably not obvious to them, since they are acting out of Instinct. In the days before marriage, these women would simply shop around the camp until they found a likely candidate, and then mate with him. Nurturing Instinct sees having and raising children as a job. They will research the details and make choices based on what will be best for their children. The problem is, the men they marry may be getting married for other reasons. And after a few years of trying to shame themselves or force themselves into making the marriage work, the ugly truth rears its head. One day she wakes up to realize she married the wrong guy for the wrong reasons. Or, the hapless lad wakes up and realizes that he is little more than a life support system for a Nurturer's desire to have a family. And at that point, things can fall apart pretty quickly.

First World - Those of us with Nurturing Instinct in technologically advanced nations sit and watch television shows about people in Third World nations and say, "Someone needs to help those poor people." We tend to see countries that are less technologically advanced as being backward and needing to work towards being more like us. We imagine the people who live in Third World nations as being depressed and starving and generally unhappy and unhealthy. Even the term "Third World Nation" is a judgment, and generally considered a negative judgment. We may believe that everyone in the world wants to be like us, or **should** want to have everything that we have. There are several tribes of humans still living like cavemen in remote parts of the world. Are they miserable and depressed? Do they long for technology? Do they have more or fewer diseases than we do? Do they live longer or shorter lives than us? Are their children generally healthier or sicker?

The booger of this Instinct is that it is so focused on it's own desires, it does not see the consequences of it's actions. It is motivated out of fear - fear of getting sick and dying. For us Nurturers to sit and watch a Third World child covered in flies and eating food with dirty hands is almost more than we can stand. We rigorously scrub our children with antibacterial soap because we believe bacteria to be bad or unhealthy, even in the face of evidence that bacteria on the skin surface actually helps to protect the skin from infection, and helps prevent colds and rashes. Like all of these Instincts, Nurturing Instinct is its own worst enemy. But, it does insure that we Nurturers will always have something to take care of.

On Today`s Show - For the first 3.5 million years, Nurturers used to sit around flickering fires and share their experience and knowledge of what was good and bad, and wrong and right. This regular gathering together and sharing was very important in making them effective in their roles as Nurturers, and keeping everyone healthy and happy.

Modern Nurturers gather around flickering televisions and watch talk shows where other Nurturers share their experience and knowledge of what is good and bad, and how to keep everyone healthy and happy. **What** we do doesn't change. **How** we do it changes constantly. We are cavemen with televisions.

The Rolodex of Known Cures - Modern Nurturers crave any information about how to cure sickness and keep people healthy. Doctors are the leaders of the Nurturers. Indeed, someone who dedicates their lives to finding cures and healing people is a hero to those of us with Nurturing Instinct. If a doctor tells us to do something to help cure a sick child, we file that suggestion away in the Rolodex of cures in our heads. Look at the explosion in bottled water sales. We never used to walk around with bottles of water wherever we went. What happened? Doctors told us that we need to drink a certain amount of water each day to stay healthy and boom! Bottled water is everywhere. We walk around sucking on water bottles as though we have been suffering from chronic dehydration for centuries, and dying by the thousands, and now, finally, we've figured out how to fix the problem. If doctors told us to eat three raw brussels sprouts a day to combat cancer, every convenience store from coast to coast would immediately stock blister packs of three brussels sprouts. Tell us Nurturers that something has a health benefit, and we will quickly turn into Gatherers and consume all we can find.

Try this: tell one of us with Nurturing Instinct that you just don't feel good, and watch us roll through the data base in our heads trying to diagnose you and prescribe a cure. "Is it in your stomach? Are you getting plenty of rest? Are you drinking orange juice? Have you tried this and that? What other symptoms do you have? Nurturing Instinct is tenacious. And if you don't get better, we will try just about everything in our Rolodex, even if it's not related to your illness. We Nurturers use cures like Warriors use weapons - throw everything you have at it and see if it fixes it.

Chicken Soup For The Cold - Some of us with Nurturing Instinct believe that chicken soup helps you to get over a cold. We may not be able to rationally explain why if you ask us, but we will recommend it every time. This is knee-jerk instinctual behavior. Our Nurturing Rolodex is usually full of things that other people have recommended to us, and we continue to pass them on. It is not our nature to question why or why not these cures are effective. **Why** do we do something is the concern of the Inventor. **What** are we going to do is the concern of the Nurturer. And **what** we will do is try everything we have learned, heard, or experienced, in order to cure sickness.

Orange juice is also a good example. It is one of the staples of us Nurturers. Anything that remotely looks like a cold gets orange juice recommended as a cure - even allergies and viruses, which are unaffected by vitamin C. Even though many other foods contains more vitamin C than orange juice, we will still recommended it. Even though vitamin C tablets are a much more efficient and effective way to get vitamin C into your system, we will still recommended orange juice. And even though vitamin C only helps to reduce the length and severity of a cold, and doesn't actually cure it, we will still recommended it. And even if you already know about orange juice, and we know that you know about orange juice, and we may even see you actually drink it, we will still recommend it. We will simply recommend drinking more. The point is, we Nurturers live by absolutes: always, all, only, and never. We will do anything we can to try to make someone well, based on the belief that, "It can't hurt, so why not?" Again, faith is a key part of our Nurturing Instinct. This we share with with Warriors and Gatherers. Inventors, Workers, and Hunters, on the other hand, believe in facts and proof. Hunters believe if you can't prove that it's actually helping, then why waste your time doing it? Workers believe everything eventually works itself out, so they don't worry about it. And where Nurturers look for the cure, Inventors look for the cause.

The Six Million Dollar Man - Modern medicine is a business, a business which generates an enormous amount of money from trying to keep as many humans alive as it can for as long as it can. Doctors even swear an oath to this. Large amounts of money and effort are put into research to try to find cures for diseases. Nurturing Instinct is very interested in finding cures for every disease so no one will ever get sick again. Medicine, including homeopathic and alternative treatments, is a business by, for, and about Nurturers. There is much talk and research about slowing and/or reversing the aging process. We are in the early stages of being able to manipulate genes and create human beings of our own design. Cloning is almost a reality. Starting with Adolf Hitler and his medical team working to create a master race, modern medicine is still trying to make a perfect human being. The ultimate goal of us Nurturers is to create a person who will never get sick and will live forever. The irony is, that if we ever do create a perfect human being that will never get sick, we will invent ourselves right out of our jobs. If we become perfect, never get sick, and live forever, why would we need Nurturing or doctors?

Immortality - We Nurturers spend a great deal of time and money trying to live forever. As Primitive people, we probably had very little time to consider such things. Most of our time would have been taken up with gathering food, finding shelter and water, and reproducing. And then we died. We probably saw death on a daily basis as we killed to eat and to survive.

Us Modern Nurturers generally do not have this connection to the earth and the cycle of life and death. And so we invest great amounts of time, effort, and money into trying to understand what there is on the other side. But for all of our efforts, no one has been able to go to the next world for a month or two, have a good look around, and come back to tell us what it is like. We don't even know if there is a next world.

There are many different religions on the planet and each one has a different idea about what happens to us when we die. However, none of these theories can be proved or disproved. And so here we sit, scared of the unknown, and back-pedaling as fast as we can.

$%&! You!** - Some of us with Nurturing Instinct believe there are good and bad words. We are uncomfortable and often afraid of certain words. What could give these words power over us? Could it be that we judge good and bad so much that nothing escapes being rated? Even words? We will avoid bad things at all costs - poisonous plants, dangerous cliffs, and wild animals. Perhaps once something is labeled as harmful, or potentially harmful, it passes around us Nurturers as an absolute. Our Nurturing Instinct sees the small picture, the immediate issue, and the task at hand. So we focus on the words, not the intentions. For example, imagine that we are doing the slippery with the person we are in love with, and in the heat of passion they whisper, "I want you to @#$% me!" Compare this to some total stranger telling you to, "Go @#$% yourself!". It may be the same word, but there is a dramatic difference in intentions. But, since Nurturers live by absolutes, we may believe that **any time **anyone** uses a certain word, it's bad. We put all the focus on the word in use at the present moment, not the situation, or the person using it, or the context it is being used in. This limits our ability to laugh, and blocks us from enjoying humor.

Is there any evidence linking the use of "profanity" to not becoming successful? Or to being sick more often? Or dying young? Or not having a functional relationship, or healthy children? It is classic behavior for us Nurturers to cling to superstition and not be able to rationally explain our behavior.

Thou Shalt Not – At the heart of what some of us Nurturers believe, there is usually some kind of faith, or organized religion. The constant task of trying to keep a child alive and healthy can be exhausting. It is helpful to imagine that a kind, loving, powerful guiding force is there to help us. But, religions are based on two things: a belief, and a moral code of behavior. Belief is very simple; there are many beliefs in the world, each with their own documents and history to back themselves up. We Nurturers will often take on the beliefs of people around us, because we see these beliefs working for other Nurturers, and that's all the proof we need. Like some other Instincts, our Nurturing Instinct will take any short cut it can find.

But where does a code of behavior come from? Our Nurturing Instinct is motivated by shame, guilt, and fear. The motivation to take better care of our young is based in guilty feelings that we are not doing a good enough job. This Instinct does not relax. There does not come a time when we don't feel concerned about the safety and health of our child. We live in fear of doing the wrong thing and having the infant die. One of us can feel shame from observing the behavior of another Nurturer, and then judge ourselves as not being as good a Nurturer. Shaming ourselves, shaming other people, and letting other people shame us, can wear us out. So, if we have some absolute guidelines to follow, then we don't have to guess, and our job gets a little easier. All of our Instincts seek to make their jobs easier. To have an absolute authority give us a clear code of behavior takes a lot of pressure off us Nurturers. The desire to do the right thing has led to written and unwritten codes that we Nurturers pass around to each other. This is where codes of behavior start.

But, with shame, guilt, and fear as motivators, these codes of behavior can develop into much broader and specific areas. "Thou shall not masturbate" and "Thou shall give 10 percent of your income to the church" have nothing to do with raising healthy children. They are more directed towards controlling peoples' behavior. But, if a kind, loving, powerful guiding force gives us a set of rules to live by, and the payoff is that our job as a Nurturer is going to be easier, we will take the whole package without question. The double bind for us is that we create codes of behavior where we feel shame about our thoughts, guilt about our behaviors, and fear of ultimate judgment and punishment. But we can only create what we know. Anything we create is simply a mirror of our selves, and invented to meet the desires of our basic primitive Instincts.

COMPATBILITY

Soul Mates - Our Nurturing Instinct may also be the origin of the idea of soul mates. A perfect match would be a very satisfying thing to our Instinct. Since we Nurturers desire absolutes, we may truly believe there is an absolutely perfect mate for us out there somewhere. This would also take the pressure off trying to find a supportive and helpful partner, and someone to have regular sex with.

Some of us will tell you that we believe we are on a predetermined path in our lives. We may attribute this to some form of creator, or the universe in general. We may talk about how everything is part of a "plan", and that we get the things we need when we need them. We may even say that there is no such thing as an accident. This is all part of the soul mate concept, and is a desire born out of our Nurturing Instinct that everything and everyone is going to be all right.

Those of us with Nurturing Instinct live according to fantasies. It is a fantasy to think that our child will never get sick and will live forever. This is the basis of our Instinct: to try to keep our child from ever getting sick, and to keep it alive forever. This is what makes us good at Nurturing. Unfortunately, what makes us good at Nurturing also makes us difficult as partners. Relationships don't exist in fantasy, they exist in reality.

And so the fantasy and the reality of our relationship will conflict constantly.

We can spend our whole lives looking for "The One" and being constantly disappointed. Meanwhile, we may overlook someone who would be a good partner for us simply because they had the wrong color eyes, or a different religion, or were a few years older than the person we always thought we should be with.

You Lost Me - Those of us with Nurturing Instinct can be confusing to have conversations with. We can perceive judgments where there aren't any, and may respond to what you are saying with an emotion which may seem to come out of nowhere. In our mind, everything is judged good or bad. This instinctual knee-jerk reaction is what makes us good at keeping children alive. But, unfortunately, for someone in a relationship with us, it can also make us confusing and frustrating partners. We may hear something that you say to us and in our mind make an assumption, then make a judgment based on that assumption, and then respond. This all takes about a fraction of a second, and then we will blurt out something angry or defensive. And you will be perplexed because you missed a couple of the steps, and are now being held accountable for an assumption and a judgment which you had no part in making.

Take the statement, "A female with Nurturing Instinct can undergo a personality change when she starts to have children. She will probably become very focused on the health, safety, and education of her children." Those of us with Nurturing Instinct may read that and respond, "So what's so **bad** about that?" Other Instincts may respond, "Yes that is true." Those of us with Nurturing Instinct can hear statements as judgments. Since we are always judging ourselves, we may assume others are doing it too. But, because of the diligent nature of our Instinct, the assumption and judgment we have made in our head will usually be negative.

Our Nurturing Instinct is always wary of unseen bad things around the next corner. The way it stays effective is to assume that we are doing it wrong, and need to stay alert for danger. And so, a completely innocent remark from you can turn into an assumption that we are doing something wrong. And, this can motivate us to judge you as being bad. And then we may respond, perhaps with some comment about a past incident where you did something wrong, in an effort to balance things out. We Nurturers tend to use comparison as a form of argument - "OK, so I maybe screwed up, but at least I'm not as bad as you!" And there you stand, blinking and confused - "OK, so how did I end up the bad guy, and why am I being dragged into a defensive position about something that I don't even know what the hell we're talking about?!?!"

We Nurturers live our lives based on absolutes: always, never, none, every, all. We will tell you straight out, "I would **never** let my child do such and such." And yet, if you say the word "Never" in a sentence, we will be the first to blurt out, "Never say never." We will tell you that we are very accepting and open minded, but our Nurturing Instinct is the second most closed-minded Instinct of the eight. Warriors are the first. A rigid code of beliefs is what makes us Nurturers good at Nurturing. But we can appear to be contradictory to others in our behavior. This contradiction causes a great deal of problems for us, and often leaves us feeling frustrated, misunderstood, and confused.

I'm JUST Saying – If you confront one of us with Nurturing Instinct about a comment you feel insulted by, you will most likely hear the phrase, "I'm just saying ..." What is happening is that you are not responding the way our Instinct believes you are supposed to, and it is confusing for us, so we try to say it again, to see if you give the correct response. And, we may think that using the word "just" actually justifies our remark. Those of us with Nurturing Instinct are unlikely to see a separation between motivation and behavior. If wey feel motivated by our Instinct to say or do something we believe will make another person's life better, then we will instinctually try to do it. You, on the receiving end, may see this as us butting in where we don't belong. In other words, we are looking at our **motivation**, and you are questioning our **behavior**. If you question us on our **behavior**, we may assume you are questioning us about our **motivation,** and feel hurt and offended, since we only wanted to be helpful.

In our mind, we **should** want to be helpful to others because it's a good and right thing to do, and you **should** want our help because we are only trying to show you that we care, and you **shouldn't** get offended because you **should** think the same way **we** do. Our

Nurturing Instinct thrives on shoulds. But, Nurturing is only one Instinct, and none of the others are concerned with the things a Nurturer is concerned with. If they were, they would not be good at their own roles in keeping people alive and well and safe on the earth.

Those of us with Nurturing Instinct see things from the point of view of how **we** would deal with it if it was **our** issue. We don't take into consideration other people's feelings, like a Worker does. Nor are we concerned about what would be the **best** thing to do, considering all the options, like a Hunter would. We offer advice without considering how you will take it, which is why we are so confused and hurt when you are offended by our advice.

Maybe If I Try Harder? - Some women stay in a dead relationship long after it has stopped meeting their needs. It is the instinctual nature of a Nurturer to keep trying to do the best she can for her offspring. Anything she can do, she will do, because she is committed to their survival. If a Nurturer has a sick child, she will do whatever she can, for as long as she can, to try to make the child well. Watch a mother bear licking her dead cub and nudging it, trying to get it to stand. She will stay with it for a long time, refusing to accept that it is dead.

This motivation can carry over into everything she does because she is driven by Instinct, not intellect. This Instinct is probably the primary reason why children have survived to become adults for the last 3.5 million years.

One downside to this Instinct is that some women might believe that since they feel this level of commitment to children and the relationship, that other people **should** feel the same level of commitment - specifically, the male they are partnered with. The problem here is; that the male they are partnered with may have Warrior Instinct, and might only feel commitment to the thing **his** Instinct drives **him** towards. This is probably why some women's best friends are other women. Maybe no one understands a Nurturer's motivation like another Nurturer.

Nice Guys Finish Last - It's absolutely true. Some women with Nurturing Instinct can tell you to the smallest detail the man they want to be with. The problem is, he is not the man she will be attracted to. The nice guy, the one who treats her with respect, the one who is emotionally present and supportive, is going to have Worker, Inventor, Nurturing, or Gathering Instincts. The guy she will be attracted to, is the one who is full of passion and self-confidence, the one who takes what he wants and charts his own course. The big strong Warrior, whose washboard abs and well developed muscular arms are seen by her Nurturing Instinct as a clear sign that this guy can protect her and her children. She can sit right next to the guy who is right for her and not even realize it, because she will not see him as a potential partner. Sometimes, she will not be able to even physically see him, because she is so focused on the Warrior across the room, passionately hot for the guy she can't stand. "Look at that guy, he's such a jerk.

But look at that chest; I just want to rip his shirt off and do him right here!" And, ironically, that is often what he is thinking too. And so it happens every weekend at a bar near you. Even if she makes a conscious decision to be with the Worker or Gatherer instead, she can become bored and end up looking elsewhere outside the relationship to fulfill her need for passion and excitement.

Again, this was probably not a problem when we lived to be only 20-25 years old, and before we invented the concept of marriage.

Therapy Sucks! - Because who are you going to date? We Nurturers invented counselling and therapists. Anything we can do to help people get along better we will. We just want everyone to be happy and healthy. And the way we become healthy and happy is to look at the reasons why we are unhappy and emotionally unhealthy. But, becoming aware of behaviors that cause problems can be a mixed blessing. First, once we start to to see our own self-destructive behavior, we also start to see it in other people. Once we become aware, and decide to try to live without destructive behavior in our life, it also becomes difficult to live with a partner who exhibits the same kind of behavior. This tends to narrow the dating pool somewhat.

Secondly, we Nurturers are highly motivated by shame, guilt, and fear. It is what keeps us focused on Nurturing. It's how we remember to floss, take vitamins, and get regular physicals. But, if we use shame, guilt and fear in our relationships, it can cause problems. Shame, guilt, and fear kills intimacy and drives people away from us. And so, awareness, for us Nurturers, can create confusion. When is shame healthy? How can we use guilt to motivate ourself, and not use it to try to force other people to self-motivate. How can we live with a healthy level of fear without it affecting our partner or children?

Partners

Nurturing and Hunting - Our Nurturing Instinct can see the Hunter as a strong protective partner and reliable mate. We will like the Hunter's loyalty, commitment level, and dedication to finding the best way to do things. The Hunter can feel supported and valued. The Hunter will probably make most of the decisions, subject to the approval of the Nurturer. This is one of the best couples for equally sharing responsibilities. Hunters are masters of organization. However, there could arise disagreements based on the best way to do things. The **best** way to do something may not always be the **good** way to do something. Hunters are realists, and Nurturers believe in fantasy.

Nurturing and Gathering - Could easily be the soul mate relationship the Gatherer is looking for. Gatherers have simple needs, and would not be a burden to the Nurturer, and the Nurturer would not have to worry about running out of basic supplies. However, both Instincts are based on fear. Gatherers fear that there won't be enough of anything, and Nurturers fear their children might get sick and die. These two could feed each other's panic about the smallest of problems, and live with a constant level of stress in their lives.

Nurturing and Warrior - Many books have been written about these two as a couple. Nurturers and Warriors themselves write lengthy volumes about what is wrong with the other one, and what they need to do to change. Many of these books make the assumption that it is a male/female issue. This is the short-sightedness of both, which believe that if **they** have a problem with something, then **everybody** has a problem with it. Both Warriors and Nurturers believe in absolutes (all, always, none, never, and every).

These are the two who will most likely be attracted to each other, because they are the ideal biological mates to produce the healthiest offspring. This is the probable origin of most of the marriages that occur, and likewise most of the divorces. Nurturers can be initially attracted to Warriors because of their strength and decisiveness. But after a while, the Nurturers will come to realize this is actually rigidity and stubbornness. Warriors talk about what is right and wrong, and Nurturers are attracted to their confidence and commitment. But after a while, the Nurturer realizes that the Warrior is saying right and wrong, and **they** are saying good and bad. These are not the same thing. Good and bad can change with new information, but right and wrong never change, even in the face of overwhelming evidence. A Nurturer and a Warrior are likely to spend a great deal of time butting heads. It is of the utmost importance for a Nurturer to have the right information. For them, its about surviving. If they do the wrong thing, their child could die. Likewise, for the Warrior, it is of the utmost importance to have all the right information. If they do the wrong thing, they could die, or the whole tribe could be wiped out. Two people who have to be right all the time may not be able to peacefully coexist. They may just drive each other crazy.

Both Nurturers and Warriors are not likely to see a difference between advice and support. Neither one of us may instinctively try to support you in finding your own answers; we will probably tell you straight out what to do. And if you feel insulted by this, we will look blankly at you and say, "I was only trying to help!" We Warriors cut to the chase: "Here's the information I think you need. Just do this, and then your problem is solved." Us Nurturers are motivated out of the best of intentions: we just want you to be all right, and to fix your problem. We Warriors are motivated out of a strong belief that we know the right thing to do in any situation. Which is why we both may be confused and feel hurt, when the other one doesn't take our advice. And then there we both stand, hurt, insulted, and confused, with nothing but the best of intentions. And, at that point, we may truly believe that men are from Mars and women are from Venus.

Neither one of us sees how our behavior affects the other one. Neither one of us sees the big picture. When we become aware of how the other one reacts, we will probably respond, "Well, you **shouldn't** feel that way, I'm just trying to be helpful!" And so we both may feel defective in our reactions, and feel shamed by the other one. And shame kills intimacy fast. We Warriors are angry and defensive by our nature. We use force, and threat of force, to get our needs met. We Nurturers use shame, guilt, and fear to get our needs met. This combination can often result in abuse and violence. If we shame a Warrior about being angry, we set ourselves up as the enemy and direct their anger towards us. And we can get hit. If we swat at bees, they will try to sting us.

There are two components to spouse abuse: anger and shame. If we shame a Warrior about their feelings, they will strike back. It is their nature. Our Nurturing Instinct will try to stop a Warrior from being angry, because we try to avoid anything negative or bad, and believe that we are only trying to be helpful. This is our nature. But neither one of us sees the big picture. Neither one of us sees the effect our behavior has on the other one. The Warrior hears the Nurturer telling them that they are wrong. Warriors believe they are **always** right. It is their Instinct. It is what makes them good Warriors, and able and willing to die to protect the ones they love. If we contradict them, **we** become wrong and, often, the enemy.

Our Nurturing Instinct has the Warrior`s best interest in mind. We want them to be happy, not angry. But, the way we do this is to tell the Warrior there is nothing to get angry about. We just want them to calm down. If we tell a Warrior there is nothing to be angry about, what the Warrior hears is: that their anger is wrong. This will almost always make them strike back at us. In their mind, we are shaming them. Shame kills intimacy as fast as abuse. And the shame doesn't have to happen every time. "My husband always over-reacts to blah blah blah." "My wife thinks I'm too stupid to understand blah blah blah." Once a Nurturer has made a judgement about a Warrior's behavior, they never forget it. So, at any time, our Warrior Instinct can fight back against a voice that a Nurturer has put inside their head.

And that Nurturer may not even be their current partner. Once a Warrior is shamed, and made to feel wrong or defective, it stays with them forever. A current partner can be abused, and suffer the retaliation of a former partner's shaming. If a Warrior is shamed by their parents, or others when they are young, they can carry that anger for life, and vent it on whoever comes close to them.

Warrior Instinct is absolute. It is unwavering in its belief about wrong and right. Nurturing Instinct is absolute. It is unwavering in its beliefs about good and bad. This combination produces troubled children, battered spouses, and murder.

Nurturing and Worker – Someone with Worker Instinct may actually be the perfect support system for a Nurturer. Nurturers just want to have a family, and keep everyone healthy and alive for as long as possible. Workers desire predictability, and intuitively exhibit strong family values and respect. There is no problem of either one feeling unimportant or not needed. The Nurturer will probably shame the Worker about certain behaviors, (Nurturers shame everyone, it's the natural outcome of constantly judging people, situations, and behavior), but this will feel normal to the Worker. Unlike a Warrior who cannot stand shame, a Worker will put up with a lot of abuse for a long time. They shrug it off, pick their fights, and look at the bigger picture. And the bigger picture for both will be: a healthy happy family. Sure there's gonna be problems every once in a while, but if most of it works, then why dump it?

Nurturing and Inventor - See Inventor and Nurturing - page 142.

Nurturing and Attraction - See Attraction and Nurturing - page 164.

Nurturing and Mating - See Mating and Nurturing - page 179.

Nurturing and Nurturing - If their values of good and bad match, this could work. But this combination can cause a variety of problems. The biggest reason for two Nurturers to hook up would be to have children. Since Nurturers typically put other people's needs before their own, these two will most likely lose their own sense of identity in the relationship. Furthermore, one person with a higher level of Nurturing Instinct than the other can cause anger and resentment that one partner is doing more than the other.

Two people with Nurturing Instinct can produce a child who is neurotic with obsessions about their health and safety, a general lack of trust, and may have difficulty separating from their parents. This could set the child up for intimacy disfunction with romantic partners and failed relationships. Contrary to their names, Nurturers are not actually the best ones to raise children. The thing that makes them effective at keeping children alive is also the thing that makes them produce children with chronic psychological and emotional problems. Nurturers obsess about the health and safety of their child.

Workers are actually better at Nurturing than Nurturers because they don't obsess over their children. So they produce more self-reliant and emotionally healthy offspring. Workers instinctually laugh and play with their children. Nurturers see raising children as a job, and one they are committed to doing as best as they can. Nurturers need to remind themselves to relax and enjoy their children's company. Nurturers talk about quality time, family night, and family values. They make issues out of things that Workers do instinctually. Where a Nurturer will make an issue out of setting aside a special day to spend time with their child, a Worker and their child simply get in the van and go.

Breeding Instincts

In a Nutshell:

Attraction - Motivates us to look and feel attractive.
Mating - Makes us wanna have sex.
Nurturing - Motivates us to reproduce
and care for our young.
It makes us constantly judge good and bad.
It makes us believe in, and live by absolutes -
all, none, every, always and never.

Breeding Instincts are satisfied when:

Attraction - We feel attractive and wanted.
Mating - We have sex.
Nurturing - We feel useful to those around us,
our children are healthy and happy.

When they are not satisfied:

Attraction - We will flirt and pout.
Mating - We can become agitated and tense.
Nurturing - We will use emotional manipulation
through shame, guilt, and fear.

Given unlimited resources:

Attraction - We will try to look 25 years old forever.
Mating - We will have sex until we can't stand up.
Nurturing - We will try to live forever.

COMBOS

Most of us appear to be some combination of these eight Instincts, and below are a few of the obvious ones.

Cheeseburger and Fries - The Worker and Gathering combo is probably the most common combination on the planet. The Worker shows up every day and does their job, whether it's punching out plastic forks in a factory or serving up hamburgers and fries. They make just enough to live on, and with a minimum of grumbling, life goes on. Gathering Instinct wants a little bit more than the Worker, and they want it now. So these people may live with a level of debt that they can just afford as long as they keep getting up and going into their miserable job. The payoff for them is the gleaming Harley in the garage that they are making payments on. It would not be unusual for people with Worker and Gathering combo to be making payments on a brand new $8000 fishing boat and be towing it with a rusted out truck worth less than $1000. It is a question of what makes them happy. The truck, which they rely on every day to take them to work so they can pay for the boat, does not make them as happy as the boat does. This combination creates the work force whose debt load keeps the economies of countries afloat. They desire consumable goods, and they manufacture them. This cycle of production and debt keeps the currency stable and the markets active.

Steak and Salad - If the Worker and Gathering combo are the blue collar segment of a society, then the Nurturing and Warrior are the white collar segment. People with Nurturing and Warrior Instincts believe there is a right way and a wrong way to raise children. They may also believe there is a right way and a wrong way to run a country, build a house, make soup, and just about anything else you can come up with. They have opinions and are not shy about sharing them. They believe in higher education and personal growth. They constantly judge other people and may often find them lacking. They may see people with Worker or Gathering Instincts as lazy, unmotivated, and poor parents. Nurturing Warriors work hard to provide better education, better housing, and better food for their families. They believe everyone else **should** be doing the same. They constantly see ways other people could be improving their lives, and will probably not hesitate to make suggestions. Since they are motivated out of what they believe to be the best of intentions, they may be confused and angry when others don't change their behavior, or react negatively to their suggestions.

Nurturing and Warrior combo is a mix of seemingly contradictory attitudes. Nurturing Instinct wants to keep everyone alive and healthy for as long as possible. Warrior Instinct is constantly alert for potential enemies and will not hesitate to kill. Both are extremes. They also live with the conflict of right and wrong, and good and bad. Right and wrong are carved in stone. Good and bad can change with new information. This internal dialog and its contradictions can manifest some confusing behaviors.

For example, white people with both Nurturing and Warrior Instincts might talk about celebrating the diversity of cultures, but then judge Hispanic people, in general, as being lazy. They may want their children to learn to speak Spanish to be more rounded in their education, but would not move their family into an Hispanic neighborhood. The Nurturing and Warrior combo doesn't usually feel comfortable living around people of ethnic backgrounds different than their own. They tend to live in neighborhoods of people similar to them in ethnicity and in income. If an Hispanic family with an income similar to theirs moves into their neighborhood, however, then they will let their children play together, and may even brag about "Their Hispanic friends".

People with this combo may vocally support improvements and funding for public education, but then send their children to private schools if they can afford it. They may support the arts, and brag about their relationships to artists and writers. They like people who are creative and expressive, as long as what they create and express does not conflict with their own political or religious views. They may support public broadcasting, opera, theater, dance, symphonies, and anything else that adds culture to their lives. But only if it adds color to **their** way of life, not if it makes them **question** their way of life. People with Nurturing and Warrior Instincts may support freedom of thought and speech, but only if it favors what they already believe. They believe that people today are much better off than we have ever been because we are striving to improve ourselves. But that improvement must be done according to what they believe is the **right** way, and the **good** way to improve.

Save the Whales - People with Nurturing Instinct are motivated to improve human beings to live longer, happier, and healthier lives. The Hunting Instinct is motivated to find the most efficient and effective ways to do things. If a person has high levels of both Nurturing and Hunting Instincts, this combination could lead them to act Nurturing on a community, national or global level.

These people may tend to be serious recyclers, shop at whole foods coops, become active in concerns about protecting the environment, and buy longer lasting import automobiles or ones that have fewer toxic emissions. Their motivation is a combination of doing anything they can to improve the human being to live the longest and healthiest life they can, and finding the most efficient way to do that. And to them, the most efficient way to extend healthy human life is to maintain a healthy environment.

This Hunting and Nurturing combo can also widen to include other species and prompt some people to become vegetarian. They may feel motivated to support organizations that strive to protect other species, wilderness areas, and wetlands.

If they also have a level of Inventor Instinct they may believe in sustainable agriculture and healing with nature. They may be motivated to live off what they grow themselves and separate themselves from others who they feel don't live in an Earth-friendly fashion. They may be advocates for solar power or wind power, use only biodegradable products, grow their own foods, and live in wilderness areas without plumbing or electricity.

If this person also has a level of Warrior Instinct in them, they can become aggressive about their beliefs and try to force other people to take better care of other humans, other animals, and the environment. They may vote for third party candidates, and be active in organizations like Greenpeace, People for the Ethical Treatment of Animals, The Sierra Club, The A.S.P.C.A., The Nature Conservancy, and Amnesty International.

BIPOLAR - The Inventor Instinct is the home of depression. People with Inventor Instinct see the biggest picture of all. They can often see life as meaningless and difficult. They believe they have no power to change the world for the better, because other people will not listen to them, and people with smaller views of reality, who are motivated out of anger and fear, are ruining the planet for everyone else. The Warrior Instinct is the one who believes they own the world and can do anything they want. If one person has both of these Instincts inside them operating at a high level, they could ride an emotional roller coaster their whole life. On any given day they could wake up and feel like conquering the world, or completely depressed because no matter what they do it won't matter anyway. Some conditions we now think are medically rooted may actually be caused by conflicting Instincts. Some forms of depression could be due to conflicting Instincts. Some Manic/Depressive conditions could be caused by conflicting Gatherer and Warrior Instincts. In fact, multiple personalities may actually be rooted in multiple Instincts.

WARRIOR COCKTAILS

WARRIOR AND GATHERING

Instincts might combine to inspire someone to stockpile weapons and supplies for the coming Armageddon. They may build bomb shelters to survive a nuclear war. They may have alternative power sources in their homes and live out in the country, far away from people who they may believe will try to come and take away their stuff.

WARRIOR AND INVENTOR Instincts can combine to make some people a threat to national security. They would see the big picture, and be willing to kill and die for it. Their views of the world may be difficult to live with, even though they may be very passionate about their beliefs. They would die for causes and leave behind manifestoes. They would have no loyalty to any flag or government. Their only loyalty would be to their own ideas. The Warrior and Inventor combination of Instincts would see things very differently than the rest of the Instincts. They would see laws as concepts, not hard and fast rules. They may get stopped for speeding and go into court and argue that they never agreed to abide by that speed limit anyway, so to fine them is pointless since they don't recognize the law as any authority. They may question the authority of a government to collect a tax they never agreed to pay. They would question everything and every kind of authority. This is how they come up with new inventions. This is also how people separate themselves from tyrannical governments and start new countries. The United States won its independence by questioning the authority of Great Britain and rejecting its taxes..

There are always a percentage of people in any society who question authority. Combine this with the forceful right and wrong nature of the Warrior Instinct and you have someone who will take the simplest matter up to a philosophical level and try to force their ideas to be accepted, even if it means a few people have to die to prove them right. As a loner, this kind of person can outwit police and military for a long time. Inventor Instinct is quick and clever. And they can build their following right under the noses of those in power, because they fly under the radar of our Warrior Instinct. We never see it coming, and by the time we catch on, we are unable to control it. If they have any charm and can get others to follow them, they lead revolutions and end up in history books. Hitler was a Warrior and Inventor combo.

Warrior and Hunting combo produces hit men, soldiers of fortune, and bounty hunters. The extreme right and wrong conviction of our Warrior Instinct, coupled with the research and detail-oriented mind of our Hunting Instinct, can create individuals who are calculating, scheming and fearless. They are extremely loyal and relentless. They would not be afraid of anybody or anything. They can be masters of strategy and cold-blooded assassins. They feel no regret or remorse for their actions because they see the bigger picture and see killing as simply a means to an end. That end can be money, protecting their country and their people, or bringing down a foreign government that they believe threatens them. These people are perfect for special operations undercover teams, bodyguards, and spys. If you add an extreme religious view into the mix, you get suicide bombers and terrorists.

Born that way?

A word must be said here about Nature and Nurture.

Being born with a certain combination of Instincts doesn't always dictate what kind of behavior we will be motivated towards. It is important to remember that how we grow up, and the behavior we witness as a child, plays an important part in our behavior as adults. Some of us grow up behaving like our parents. Some of us grow up with one or two similar behaviors, and some of us are vastly different than our parents. The difference is not only our Instincts, but the behavior we observe our parents doing, and the behaviors our parents try to force us to do.

If we have Gathering Instinct and live with a workaholic father, one thing may feed the other and we could also become a workaholic. Likewise, if we have Gathering Instinct, and grow up in a family that has no addictive behavior, we may become a pack rat. Or, once we reach adulthood and discover the power of money, we could turn into a workaholic based solely on the motivation of our Gathering Instinct.

Those of us who receive little affection and support as children may grow up believing there is not enough of anything in the world. If we have Gathering Instinct we may feel driven to gather whatever we can to fill the void: food, money, lovers, and possessions. Again, the behaviors may be similar, but the difference lies in our motivation. If you ask **why**, you will encounter our belief system, and, subsequently, our Instincts.

Those of us who grow up with an overbearing parent with Warrior Instinct may learn to behave out of our Warrior Instinct, if we have it. But, another child in the same family who has Inventor Instinct may internalize it and become depressed and sullen. The message we receive is that we are supposed to act like a Warrior, but if we don't seem to be instinctually motivated to be aggressive, we may judge ourselves as being defective. This can also be mirrored back to us by our parent: "Why don't you try harder? Why can't you be more like your sister?" To those of us with Inventor Instinct this can reinforce what we instinctually believe: that we don't fit in with the rest of society, and our lot in life is to be the odd ball misfit.

Those of us who are children of emotionally healthy parents who feel loved, supported, and encouraged to be ourselves, can develop behavior based solely on the balance of Instincts we were born with. If we have Inventor and Nurturing Instincts, we could grow up to find a cure for cancer and save the whole world. But, if we have Warrior and Worker Instinct, we can still become depressed and ride a roller coaster of emotional ups and downs our whole lives, despite the love and care of our parents.

Likewise, if we are the children of emotionally or physically abusive parents, and if we have Warrior Instinct, we can grow up fighting and rejecting our parents. By acting on the determination to be nothing like them, who we see as wrong, we can do great things and become a rags-to-riches hero.

The point is, these Instincts are already inside us, external stimulus simply brings them out.

Take, for example, two girls with the Warrior and Nurturing combo. One grows up with hippie parents, she shops at whole foods coops, wears clothing made from hemp, supports the Sierra Club, and marches in protests. She lives in a rural area with solar power and learns how to grow food and recycle. The other girl grows up in a 1.7 million dollar house with parents who are attorneys. She attends the best schools her parents can afford. She gets a credit card and a car for graduation, and learns how to dress for success, and advance in the business world. Both girls will live their lives with clear and strong convictions about what is right and wrong, and what is good and bad. *How* they live their lives may be polar opposite. Their beliefs may differ, but their attitudes and their behavior will be similar. They may both argue with strong conviction against the other one's beliefs, but instinctually, they are more alike each other, than they are different.

It is our Instincts which bind us together. It is also our Instincts which separate us. And often, it is the same Instinct which does both.

Bad News & Good News

THE BAD NEWS

If it is true that these behaviors are motivated out of basic primitive Instincts, then we may not be able to change them. No matter how much professional help we get, we may not be able to change the way we act. Instincts go against common sense and rational thinking. Remember, there are three kinds of behavior, learned, forced, and instinctual. We can learn new behavior, and it can become second nature in certain situations. We can force ourselves or stop forcing ourselves to do certain behaviors. But, situations are going to arise which trigger instinctual behavior, and we are going to react without thought or control.

We think we can control things. But, the systems we invent, to make our lives easier, continually fail us. Of course they do. Why? Usually only one Instinct is involved in the design. We design technology, social structure, laws and systems to satisfy the desires of our **Instincts**, not to meet our **needs**. Human beings have four basic needs: water, food, shelter, and safety. But, we have yet to structure a nation around meeting these four needs. Never has there been a government divided into four sections: The Department of Water, The Department of Food, The Department of Shelter, and The Department of Safety, where each of these departments is responsible for seeing that every human in their country has these four basic needs met, and the tax money is split evenly between them. No, governments are designed, and countries are structured on satisfying the desires of our **Instincts** instead. Or, more specifically, the Warrior, Nurturing, and Gathering Instincts. Remember, these 3 live with a level of constant vigilance and fear, which makes them good at their jobs. But their fear is infectious and they can scare the rest into letting them design and run things.

After the stock market crash in 1929, did we learn our lesson and separate our needs for food, water, shelter, and security, from being affected by the stock market? No, we went right back to the same system which failed us. We are like the battered wife who keeps going back into the abusive relationship, each time expecting it to be different. Why? Fear. We are afraid of the unknown more than we are afraid of a known evil.

Our Warrior Instinct lives in fear of other people stealing our resources. Our Nurturing Instinct lives in fear of us getting sick and dying. Our Gathering Instinct worries and panics about not having enough of anything. In some countries, the most profitable segments of society are the defense industry, hospitals and drug manufacturing, large corporate chain stores that sell inexpensive consumable goods, and fast food. The other most profitable business is insurance. This is a business which makes money from selling peace of mind to people with Warrior, Nurturing and Gathering Instincts. We even have laws that force citizens to buy insurance. Civilization itself is structured around the fears and desires of these three Instincts.

We rarely use our Hunting, Worker, and Inventor Instincts in the design and operation of a country. Consequently, a portion of the population is satisfied, but many others are not. Those Instincts which are satisfied shame the others: "Well it works for me! What's your problem?" Their problem is, that they have different Instincts. What satisfies our Worker Instinct doesn't necessarily satisfy our Hunting Instinct. We haven't figured out who to put in charge of what yet. We send our Warrior Instinct in to negotiate peace. We put our Gathering Instinct in charge of efficiency. If there is a problem we point fingers and expect people to change. But people cannot change their basic Instincts.

We cannot "fix" an Instinct by forcing it to be more like another Instinct. Why are so many prisons and other correctional facilities unable to correct the behavior of some inmates? The current popular theory is that there are some people who are just bad natured, perhaps born that way or conditioned by society and their childhood experiences to grow up doing bad things. The truth may be that they are simply acting out of their Primitive Instincts. The man in the street who takes someone else's money by force, gets arrested and punished for being a bad person. But if he does it in a corporate boardroom, he gets praised as being a captain of industry. Our value system is uneven. But Instincts know nothing of values. No matter how severe the punishment is, people still commit crimes. You can get the death penalty for killing someone, but people still kill people every day. Instinctual behavior cannot be "corrected".

Instinctual behavior is also relentless. We are never satisfied. We live two to three times longer than we ever have, we have hospitals and dentists and drug stores, we have hot and cold water delivered right to our sinks, we don't have to poop in the bushes anymore, we don't have to trap and skin a rabbit just to eat, in one hour we can drive a distance that used to take us a whole day to walk, and we can pull out a device half the size of our hand and push a few buttons on it, and talk to someone on the other side of the world. You would think we would be satisfied. Walking around with big smiles, all happy for the advances we've made. But no. We rush around panicked and obsessing about potential problems. We have all the things we need to be happy, but we aren't all that happy are we? We are panicked about dying, weighing too much, having yellow crooked teeth, running out of toilet paper, and other people stealing our stuff. We have all the resources of the world at our disposal and yet, we gather like addicts and fight like there is only one loaf of bread left on the earth and we all want it. Why? We have the ability to have food available everywhere at any time and yet we still eat like we don't know where our next meal is coming from.

The more we are able to manipulate our environment and provide our selves with unlimited resources, the more obsessive we become. The Instincts which look at the small picture, start looking at smaller and smaller pictures. They get more focused on details, and obsess about smaller and smaller issues. The Instincts which focus on the bigger picture start focusing on bigger and bigger pictures. They lose touch with reality through their ever widening focus. Our fears are still our strongest motivators.

Our Warrior Instinct defines freedom as living free from fear. And the way we live free from fear, is to control, or get rid of the people and things we are afraid of. Our Inventor Instinct defines freedom as living free from **any** kind of control whatsoever. Especially the control of our Warrior Instinct. How can these two Instincts live peacefully together in the same country, or even inside the same person? They can't. We have the ability to live peacefully and content with all of our needs met, and yet still we fight. Why? Because our Warrior Instinct has nothing to do. Peace doesn't last because we have an instinct that keeps rising up periodically looking for an enemy. Now, some of us are going around the world looking for **potential** enemies. We are determined to find people who **might** threaten us down the road, and kill them. We are listening in on our neighbor's phone conversations to see if they think and act the same way we do, and reporting them if they don't. Our Warrior Instinct is creating its own enemies, because it needs them to feel useful. Long periods of peace don't work for our Warrior Instinct.

So this is really bad news: We're stuck this way. We cannot change our basic Instincts, and they cause us problems. So are we doomed? Do we have **any** power over our primitive Instincts?

One way to guarantee no more conflict would be to kill all the people with Warrior Instinct. If you have no Warriors, you have no conflict. Of course we would never do this. Another way would be to shut down all systems of health care. Close the hospitals and clinics, and stop making medicines. Do away with doctors and go back to living only 25 years. This would dramatically decrease the population. Consequently, we would not get in each other's way so much and have to fight over resources. Of course, we would never do this either.

If we continue living the way we are, and working to satisfy our Instincts by providing them with unlimited resources, our Gathering Instinct will gather until there is nothing left to gather. Our Nurturing Instinct will continue to have children until there are more people than the ecosystem can support. Our Warrior Instinct will try to kill everyone we perceive as an enemy until we are the only ones left. It is the short-sighted Instincts which may be our demise. And in so doing they may bring the human population to the point of near extinction. The Instincts which see the bigger picture would never let this happen. And so, many of our political struggles are centered around our Instincts that see the bigger picture, trying to not let our Instincts with the smaller view, destroy our ability to live on the earth. But, our Warrior, Gathering, and

Nurturing Instincts have more power and persuasion, and the short view seems to be in charge more often. The balance between these Instincts is something we have difficulty trying to control.

Control is an illusion. We can barely control our own behavior, let alone the behavior of other people. Years of standing around pointing our fingers at other people and telling them how they **should** be acting has changed nothing. We can't even stop the guy with Warrior Instinct who lives next door, from telling us how we **should** mow our lawn. And we may not ever be able to. If these Instincts are the things that motivate us, then they will continue to do so as long as there are human beings. And that's the bad news.

But then again, there is the outside chance that we could evolve out of them in a few more million years.

The Good News

Hey, we're not so screwed up after all!

Some of our behaviors are just knee-jerk reactions to certain situations. We don't intentionally mean to cause problems. Our motivation is purely to help ourselves and each other to survive. Our behaviors may be confusing , but by learning what our motivation is, we can learn to separate our motivation from our behavior. And then we can understand each other a little clearer. And not take things so personally.

Wisdom is power. We can change. Look at our history. Not so very long ago, we believed the earth was flat and we might sail off the edge. As recently as 80 years ago, many of us in the United States believed that women and black people were too stupid to vote. And now we know that we were all wrong.

There are only three things we cannot change: our race, our age, and the fact that eventually we will die. Everything else is optional.

Perhaps, we can learn to live with each other a little bit better if we understand the motivation behind certain behaviors. Perhaps we can try to see people in a new way, and not take their behavior as a personal attack or some kind of judgment on us. If we can figure out the strengths and weaknesses of each Instinct, then perhaps we can channel them into areas where they can serve the greatest good. Put those of us with Hunting Instinct in charge of budgets, and put those of us with Inventor Instinct in charge of making policy. If we can figure out who to put in charge of what, things can get taken care of, by people who are motivated to take care of them. Maybe we can be a little more accepting of ourselves and a little more tolerant of others. Perhaps, just perhaps, if we come to understand our inner Instincts and how they motivate us, we can use this knowledge to create safer, saner societies to live in. If we stop letting the anger and insecurity of our Warrior Instinct define our nations, if we stop feeding the anxiety and panic of our Gathering Instinct, and if we stop clipping our own wings with the guilt and shame of our Nurturing Instinct, then maybe we can live as happy healthy humans on the face of this incredible planet. All of this is within our ability to do.

The way we change a society, is to change ourselves because we are society.

We need to learn how to separate our **needs** from our **desires**. We need to set up systems where our basic needs for food, water, shelter and safety, have nothing to do with the financial stability of corporations. It is foolish and short-sighted to base our ability to feed our children, on the stability of the stock market. We need to put Instincts which see the bigger picture in charge of such important things. We have been alive for almost four million years. We have been civilized for only about ten thousand. And, we have only been able to manipulate our natural resources on a large scale, for about the last two hundred years. We are still learning. We learn by trial and error. And now, we can learn from our experiences, how to use our Instincts to benefit us best.

If these Instincts **are** real, and they are motivating us to do these behaviors, then maybe, just maybe, by studying them we can learn ways to make sure **all** our needs are met, and **all** our Instincts can have constructive outlets. In this way, we might be able to effectively resolve some of our differences and secure our needs, and the needs of our fellow human beings.

It's just an idea. What do you think?

ACKNOWLEDGEMENTS

I would like to thank:

Garuth - you saw me at my worst, and stood by me when no one else would, I love you and treasure your presence in my life. This book would not exist without you.

Nic - you suffered through the worst part of writing this: watching me obsess over fonts, screwed up software, and broken printers, and yet you patiently let me explain the same things to you over and over. Why you didn't run away is beyond me. Thank you, my sweet love.

Thank you to Dr. Goodman and Dick, and Dave for saving my life. Thank you to Kelly and Martha for teaching me how to love. Thank you to Holly and Molly for making me feel normal. Thank you to my excellent friends Baci, Bob, Wendy, Heidi, Todd, Tim, Brad, Jim, Jimi, Dr. Tim, Cord, Dave, Jason, Bob, Anne, Becca, Lori, Steph, Bob, Marilyn and Mike, Julie and Drew, David, Kyle, Mike, Jeanine, Michael and Lauren, Dave, Laura, Swanny, Eric, Patty, John, Chris, Deb, Leah, Larry and Molly, and the Fiksdal clan. Thank you to my pals Tony and Jason for your invaluable professional insight.

Thank you to all the people whose houses I painted. You gave me the opportunity to do some fun creative things, you shared yourselves and your families with me, and you supported my work. And for all of it I am sincerely grateful

This book is dedicated to my father Reginald Gray.

You are my greatest inspiration and the biggest thorn in my side. You are my role model for success, and for failure. And now I can see how your Instincts blocked you from being a father. I have learned many lessons from you. You taught me how to learn. You taught me self discipline and perseverance. You taught me to work hard and find my own answers.

You also taught me how to self destruct and chase people away from me. And you taught me how to lie to myself, and how to be emotionally distant. But the most important thing I learned from you, was to not waste my life the way you wasted yours. I thank you for the good stuff. And I give you back the bad stuff. I don't need it anymore. Thank you, Reg, I wish you were here to read this.